Cosmologists—often in error,
never in doubt

— Lev Landau

Before the Big Bang

Before the Big Bang

The Origin of the Universe and What Lies Beyond

Laura Mersini-Houghton

MARINER BOOKS

Boston   New York

HarperCollins books may be purchased for educational, business, or sales promotional use. For information, please email the Special Markets Department at SPsales@harpercollins.com.

FIRST EDITION

Designed by Chloe Foster

Library of Congress Cataloging-in-Publication Data has been applied for.

ISBN 978-1-328-55711-7

22 23 24 25 26 LSC 10 9 8 7 6 5 4 3 2 1

TO MY DAUGHTER, GRACE,
and
TO MY DAD

The gates of hell are open night and day;
Smooth the descent, and easy is the way:
But to return, and view the cheerful skies,
In this the task and mighty labor lies.

   —Virgil, *The Aeneid*

Contents

Prologue: My Albanian Universe

A lbania, the land of my birth, is a beautiful country along the coast of the Adriatic Sea with the kindest people but the cruelest past. Home to fewer than three million people, the country is just over eleven thousand square miles, only slightly larger than the state of New Jersey and about half the size of Ireland. Yet during my childhood, its totalitarian government wanted us to believe that we Albanians resided at the center of the cosmos.

Until 1991, when Albania's Communist regime joined its erstwhile ally the Soviet Union in the dustbin of history, my country was the North Korea of Europe: poor, paranoid, and cut off from contact with the rest of the world. Peering beyond our barbed-wire borders was forbidden by an all-powerful government that crushed dissent with internal exile, hard labor, and the death penalty. The state maintained hundreds of thousands of concrete bunkers scattered across the country to protect itself against the "Anglo-American threat"; it imprisoned and persecuted many thousands of Albanians in camps modeled on Joseph Stalin's harsh Soviet gulags. Foreigners were not welcome, and Albanians were not allowed to travel abroad. Inside the country, the government decided where and how you could live. Displays of individuality were punished.

The only place that was open to us was the sky and the stars above. The state could not prevent us from looking up. Even as a young child, I could escape into the sky.

By sheer luck, I had an opportunity for escape that others didn't. When I was growing up, my mother worked at the Albanian League of Writers and Artists, an organization for artists, writers, and composers. Her workplace had its own special library where I could access books in English, which the state had classified as prohibited Western literature, forbidden to ordinary Albanians. Through these books I traveled to the ends of the world. But I could not share my dreams and imaginings with anyone beyond my parents.

My father anticipated the frustration that boundaries breed in a curious child, so he devised clever ways of using books, art, and music to channel my curiosity and strengthen my mental resilience. Classical music became our encrypted language, a code through which we could escape our everyday circumstances and share our contemplations of the beauty of the universe.

In the eyes of the Albanian Communist government, my father had a "bad biography"—his family had been landowners for generations in pre-Communist times, and because of this, they were continuously persecuted. My grandmother did not see her brothers until 1991; one of them was in jail for nearly fifty years. Several other relatives were forced into camps, shot, or exiled. My dad's first cousin, an engineer who was part of the project to drain the swamps in Albania after World War II, was taken from his home in the 1970s, shot dead, and carted off. Two decades later, his family found his body at the medical school in the capital city, Tirana, where it was perfectly preserved in formaldehyde and being used to teach anatomy. By then, the dead man's brothers and cousins were in their seventies, but the corpse they found remained the image of its thirtysomething self. The man finally received a proper burial in 1997.

My dad fared better than his cousin; he was temporarily exiled. This happened several times over the course of my childhood. His first exile occurred when I was five years old and was triggered by a letter.

At the time, my father was a professor of econometrics at the University of Tirana. He had been working on a difficult mathematical problem that had important applications from economy to astronomy. It involved the inverse of very big and very sparse matrices, which resemble massive spreadsheets with hundreds of rows and columns but where most of the entries are zero. This work earned him an invitation from the University of Oxford in Great Britain—he was offered a six-month sabbatical to discuss his new algorithm. But the letter never reached him. It was intercepted by the Albanian government.

Instead of visiting Oxford, my father began his exile at the start of the academic year, my first day of school. That morning he took me to school in my new uniform, holding flowers for the teacher, as if all was well. But because I had overheard my parents' whispered conversations, I knew he would not be collecting me from school that afternoon or for many days after. Nevertheless, I pretended to be excited so that the hopelessness I felt would not be the last image of me my father took with him to exile.

My mother and I had already lived through much drama in the previous months, a dark time during which my father underwent a process that was euphemistically called an "ideological debate" but that was in fact a trial that would decide his fate and punishment. In those days, there were no defense attorneys or real trials in Albania; the purpose of the "debate" was to have fellow professors and university employees demonstrate to my father that receiving an invitation from Oxford meant he had erred ideologically and as a result, he needed to be punished and rehabilitated. This sham went on every day for two weeks, often until midnight. At the end of the debate, a decision would be made on his punishment, which could be anything from jail, to exile, to the death penalty.

We didn't know how long the debate would last or what decision would be made. My mother and I waited every night with the lights switched off and our faces pressed against the window, hoping to hear my father's footsteps before the next tick of the wall clock; we were not allowed to openly show affection to someone who was under investigation by the party. But the memory of the terror we experienced before we finally heard the sound of his footsteps and felt the warmth of his embrace after he closed the front door continues to chill me to my core.

On that first day of school, kids gathered in the schoolyard. One by one, their names were called, and they were sent to smaller groups and introduced to their teachers. My new teacher's name was Shpresa, which means "hope" in Albanian. I pretended to talk with the other kids after I handed the flowers to my teacher, but I was following my father out of the corner of my eye as he slowly receded from the school fence. I did not dare turn my head. I remember his look of loss and profound melancholy as he glanced back one more time before rounding the corner and vanishing. I did not know when or if I would see him again.

Eventually, my father was allowed to return home for one weekend a month. (Finally, after two years, the government realized they needed his expertise again, so he was released and reinstated at the Academy of Science.) It was during his brief visits that he taught me some of his most enduring lessons—late-night wisdom that I would carry with me on my own journey through the darkness.

In those days, Radio Tirana would broadcast an hour-long classical music program at 11:00 p.m. each Saturday. During his visits, ignoring my mom's protests, my dad would wake me just before the show began. In the silence of the night, we listened together to the divine notes trickling into the living room from what seemed like the other side of the universe, carried from another space and another time. His whispered commentaries over the music and during these interludes fueled my own lifelong admiration for human ingenuity and achievement.

On one of those Saturday nights, the selection was Bach. His Toccata and Fugue was my dad's favorite piece, though the Brandenburg Concerto No. 3 became my own. "Shh! Listen to this part," he would say. "It sits at the depth of human condition and misery. It rocks you gently as it rolls down, then farther down, and yet farther until it reaches the bottom. But listen . . . and you can hear—Bach is not disturbed. He is totally serene even in his darkest moments. Can you feel the tranquility in his sadness? He is all too familiar with the human condition and the contrasting emotions of anguish and happiness. He knows the sorrows, like the highs, are part of life.

"He is peaceful," my father continued, "because he is aware of the root of his problem: His music exists outside of time, and his genius is ahead of his time. But he has made his choice. He has chosen to raise the bar higher, to produce beauty for eternity without regard for what might please contemporary audiences. When composing his works, he is driven by a higher calling, one that comes from within."

Then onto the allegro. "There now, listen . . . listen, can you hear that?" Suddenly the notes were rising again, gently carrying away the waves of bleakness.

After the music was over, my dad would explain, "You see, Bach knows he has created something special. Many countries had kings, queens, dictators at one time or another in their history. Bach, Beethoven, and many other great composers, mathematicians, and scientists worked for those rulers. But they were not confined by their circumstances. They were driven by their own passion to produce masterpieces.

"Their work made them free. For it is only books, art, music, and discovery that elevates us into truly free human beings. You too can choose to surround yourself with this treasure of knowledge and creativity."

My dad and I were not musicians, but we found scientific inquiry as inspiring as Bach's music. Indeed, these whispered conversations provided some of the earliest inspiration for my own scientific journey. There were other early influences too. Perhaps the indoctrination I witnessed during my Albanian childhood chiseled into my personality a desire to seek answers for myself by applying logic and tests, even if sometimes that meant going against established beliefs. But that curiosity took me on an odyssey richer than either my father or I could have imagined: a quest for the underlying workings and mathematically encrypted beauty of the universe.

Today, I live in the United States and am a professor of theoretical physics and cosmology at the University of North Carolina. The universe's origin is now a central topic of cosmology and one of my main areas of study. And I can seek answers, because, thankfully, in Western academia, no question is forbidden, not even the two greatest cosmological questions of our time: What is the origin of our universe? And what lies beyond?

Since my university days, I have been fascinated by the first of these questions, one that is ages old: If our universe has not always existed, how did it originate? This, in time, led me to ask follow-up questions: What was in place before our universe's birth, and what is beyond its edges?

These questions were followed by the most radical question of all: Are we simply one universe, a cosmic Albania where the world ends at our borders, or are we part of a larger cosmos, home to many universes—a multiverse, of which our universe is simply one humble member?

Using the latest advances in theoretical physics, I developed and pioneered a theory to help explain the universe's creation. For the first time, this theory provided an answer to the question of our unlikely origin. But it also went further, offering a glimpse of the vast multiverse in which our own universe sits.

Central to my theory is the notion that we are part of a multiverse—that there are other universes beyond our own. To many critics, the concept of a multiverse is purely speculative, a flight of theoretical fancy that can never be tested and is thus scientifically useless. But as a result of my theory, we have shown the opposite to be true.

In the early 2000s, by applying the laws of quantum physics (such as cross talk between universes in the multiverse due to quantum entanglement) to the problem of the origin of the universe, my collaborators and I derived a series of predictions from our theory. Our predictions demonstrated how we could glimpse the world beyond the borders of our own universe and find its fingerprints engraved right here on our sky.

Taken together, our theory and its testable predictions showed that the answer to the origin of our universe can be scientifically derived, and the existence of the multiverse is in fact testable. Of course, these tests necessarily rely on indirect evidence rather than direct proof, because we will never be able to travel beyond the point of no return—the horizon of our universe, the distance from which not even light can reach us—to obtain the kind of incontrovertible evidence that would satisfy every skeptic. But working with the evidence available to us, we can still learn an awful lot about the birth of the cosmos. And the fact remains that nearly all of the anomalies we predicted have now been observed in our distant skies.

The idea that we live not in a single universe but in a multiverse has been contemplated by philosophers since antiquity. Since the earliest civilizations, human beings have wondered how the universe started, how it will end, and what, if anything, might lie beyond. Many of humanity's fundamental questions about the universe have changed relatively little over the millennia.

The possibility of many worlds was introduced into Western thought through the influence of the atomists in ancient Greece. Atomists thought of the world as made up of indivisible clumps of matter (atoms) and empty space (voids) through which those atoms moved. In their view, a collection of atoms moving around the voids clumped together to form larger objects, such as stars, planets, and then the whole universe. Because there were an infinite number of atoms and voids, this process could be continuously repeated to form many universes.

The main difference—and it is a crucial one—between these early thinkers and current scientists is that, in the past few centuries, our accumulated knowledge about the theories of nature and technological progress have allowed us to pursue scientific investigations and subject them to observational tests for what had once been purely philosophical ideas. Scientists can now derive and test what previous generations could only imagine.

What we are finding from these theoretical and observational advances is poised to upend centuries of mainstream thinking.

Our results also challenge the dream, long cherished by physicists, of discovering a blueprint for a cosmos that contains only a single universe—a dream that entranced many of the greatest minds in theoretical physics in the twentieth and early twenty-first centuries, Albert Einstein among them.

Thanks in part to our work, the idea that our universe is not unique but belongs to a much larger cosmic family—the multiverse—has recently moved from the fringes of cosmology to the scientific mainstream. How that happened and how the idea of the multiverse came to be embraced is the story at the heart of this book.

Why does the origin of our universe matter? Truth be told, many scientists research the universe and its origin out of simple curiosity, not an expectation of immediate, practical applications. To use a distinction that many of these researchers themselves employ, the origins of our universe have traditionally been the purview of pure science rather than applied science.

Evolution has trained humans for the pursuit of science. We possess special traits, such as childlike curiosity and an innate desire to understand our environment, which have led our species to develop larger brains compared to other inhabitants of this planet. These qualities are representations of something known as neoteny, the phenomenon whereby adults retain these characteristics throughout their lives.

Pure science and applied science have different focuses, but they share a powerful symbiotic relationship. Applied science cannot exist without the discoveries made by pure science. And although proponents of pure science may sometimes show disdain for applied science and its practical applications, history has demonstrated that pure science inevitably leads to practical applications that can and do transform our lives.

There is a true story about Michael Faraday, the nineteenth-century scientist who helped unlock the mysteries of electro-magnetism: The British chancellor of the exchequer (the country's equivalent of a minister of finance) visited Faraday's lab one day. At the end of the visit, the chancellor said, "This is all so incredibly impressive, but what is it good for?" Faraday replied: "I don't know, sir, but I am sure you will tax it one day." Indeed, today, billions of humans on the planet pay for electricity and cannot do without it.

If anyone had asked Einstein what his theory of relativity was good for, he might have given the same answer as Faraday. Many of our increasingly indispensable high-tech products, such as GPS devices, are based on Einstein's work. Modern neuroscience's mapping of the brain and the electronic trading programs that govern the stock market operate by the same set of quantum principles and rules that Einstein used to explain the motion of the planets and the speed of light in the universe. Human curiosity about how the stars in the sky shine has given us the tools to produce nuclear energy, nuclear medicine, and (unfortunately) nuclear weapons. Understanding stars and structure formation in the universe led to fusion and fission potentially providing green energy here on earth. Internet, Wi-Fi, computing, and all the electronic gadgets that we now depend on, as well as ATMs and wireless bank transfers, medical imaging machines, and modern medical equipment, would not exist without the theory of quantum mechanics that Einstein and his theoretical-science contemporaries helped create.

Someday, we might derive similar benefits from discoveries related to the investigation of the multiverse. Who knows what technological advances might be unlocked if we had a better understanding of our universe's origins? Who knows what ingenuity and creativity might be unleashed if we allowed our minds to accept a premise that defies centuries of scientific orthodoxy? As we learn more about the true workings of our cosmos, we will find that our greatest scientific breakthroughs and discoveries lie ahead of us.

Being a part of this scientific quest is both daunting and uplifting. But above all else, it is an inspiring process—one that I aim to share with you now. In the pages ahead, I will describe my personal journey through the wonders of the cosmos to seek an answer for our origins and search for evidence of our vaster cosmic family, the multiverse. Just as we once overturned the belief that Earth was the center point of the universe, with the sun, moon, planets, and stars all orbiting our planetary home, now we are demoting our universe from its historic place at the center of the cosmos. In so doing, we are rewriting the story of our own origins.

Before the Big Bang

Is Our Universe Special?

O<small>VER THE YEARS</small>, I have seen my share of airports around the world, but Tirana's Mother Teresa Airport looms the largest in my memory. It was the place where my scientific journey literally took wing.

In January 1994, I left my walled-off life in Albania for a new one in the United States. My journey followed a route I never could have imagined: I had won a full scholarship to study at an American university.

Several years prior, as Albania had begun to shake off the chains of Communism, the U.S. embassy and the American Cultural Center reopened their doors, which had been closed for almost fifty years. Soon thereafter, the United States offered its Fulbright Scholar Program to Albanians. I was preparing to graduate from the University of Tirana with a bachelor's degree in physics and was deciding what to do next with both my life and career; I knew that I wanted to pursue an advanced degree, but Albania's educational system didn't offer a graduate physics program. Studying abroad was the only route.

The Fulbright program was widely advertised, and students were encouraged to apply. Based on my experience in Albania, where merit carried very little weight, and political connections were everything, I thought the idea that someone like me could receive a scholarship to study in the United States by filling out an application and taking some standardized tests was too good to be true. Urged by my friends, I completed the forms, but without much hope, so I was shocked when, a few months later, I received a letter congratulating me on winning a Fulbright scholarship to study advanced physics for one academic year at the University of Maryland in College Park. It was the first Fulbright awarded to an Albanian for science. It was also my first experience with the fairness of the merit-based U.S. system.

My parents and brother accompanied me to the airport to say goodbye. Besides math and family, my dad's two other passions were mountain climbing and chess. As my mom's and my brother's eyes welled with tears, my dad couldn't resist offering some last-minute advice. "Really good science is like climbing a mountain," he said. "It takes superb skill, stamina, and the courage to never compromise your scientific integrity. But the view from the top is breathtaking and worth the sacrifice." He added that, as in chess, "If you want to get it right, you must be at least three moves ahead of your opponent. You have to anticipate and prepare for what can go wrong, not only on the first move, but also on any of the possible combinations and outcomes of the second and third moves." By this, he meant I should thoroughly scrutinize my ideas and whether they agreed or disagreed with the bigger picture so I could anticipate where they might go wrong.

As we hugged, I started to miss them already. My dad pulled back and looked at me, and in a shaky voice he told me what only parents are capable of saying to their children: "Do not look back. We will be fine."

I boarded a Swissair plane, wondering what my new life would be like in a country where I knew no one, a sea and an ocean away from my childhood home. I could not have predicted that my one-year Fulbright would lead to acceptance to an American graduate school and, from there, to what has become a lifetime of study. Nor could I have guessed that the United States would become my new home.

Mist enveloped the plane and obscured the ground as we landed at Baltimore/Washington International Airport. I passed through customs and found someone from the University of Maryland office for International Student and Scholar Services waiting to greet me. Outside, large snowflakes were falling steadily. By the time we arrived at a hotel near the campus, the snow had turned gritty, and the storm transformed into a historic blizzard. All the airports, the city of Washington, DC, everything came to a standstill. As evening approached, I watched the fast-falling snow and icy roads as they glimmered eerily under the streetlights.

I spent my first week in the United States existing on dough-nuts and filtered coffee that the hotel kindly provided in the lobby for all its stranded guests. The interwoven moments of quiet contemplation and wonder I experienced that first week (along with the sweetness provided by all those doughnuts!) would become a good metaphor for my new life in America.

Many things appear obvious in retrospect, but truth be told, I almost didn't become a physicist. As a high school student in Albania, I hadn't been able to decide whether to major in physics or math at the University of Tirana until a week before the selection deadline. I liked them both. I competed in the national Olympiads in math and in physics, hoping that would help solve my dilemma. As it happened, I won both, which made my choice even harder. So I flipped a coin: heads, physics; tails, math. It came up heads.

But although chance had landed me on the path of physics, I always knew I wanted to be in the world of numbers and hard sciences. At that time in Albania, the fields of social sciences, economics, and the humanities were mostly political dogma in disguise. History was our national version of Communist myths and fairy tales, and even the law school was just a name because there were no defense lawyers. For me, those fields held no temptation. Yet many of the students who were assigned to the natural sciences saw it as a punishment. The math and physics building was mockingly known as "the Winter Palace."

I loved math, however, because its pure logic and precision removed all ambiguity and arbitrariness, a rare quality in Albanian life. I loved physics just as much because it combined math with creativity and intuition and, through ideas, applied math in a real setting. I ended up majoring in advanced physics in what was known as the five-year program; in the second year, I decided to sign up for the math program too.

Having a math degree in addition to a physics degree would have made absolutely no practical difference if, as seemed likely at that point, I spent my working life in Albania. But nevertheless, my parents supported me in my decision. I think my mom was happy that I was going to be fully occupied studying for two degrees and would have no time for partying; my dad, for his part, was delighted that I shared his passion for math.

My friends in Albania thought that doing math for fun was nuts, but once I migrated from the hotel to a small apartment and began navigating the sprawling campus at UMD, I found myself surrounded by students who were as passionate as I was about the subject and who were similarly determined to make the most of the educational opportunities available to them.

The University of Maryland physics department has a big graduate program. It offers an unusually large number of advanced courses in physics, along with many research programs, including world-class research groups in theoretical physics. During my Fulbright, I took full advantage of these opportunities and signed up for many more classes than required. I also applied to UMD's graduate school so I could continue my studies after my fellowship ended. Fortunately, I was accepted.

Of the roughly two hundred physics graduate students at the University of Maryland, only three were women. But other than the severe gender gap, there was a great deal of diversity. For nearly all of my life in Albania, until the regime fell, I had seen only other Albanians; to be surrounded by students from so many backgrounds and places was a new and wonderful experience. Clearly, the world was bigger than I had seen.

By my second year at Maryland, I realized that I was drawn to the big questions about the universe, specifically to theoretical physics and cosmology, the study of the whole universe—a discipline of literally cosmic proportions.

In their working lives, theoretical physicists aim to decipher how nature functions, from the smallest conceivable particle to the largest distances. We do so by working with the current laws of nature and discovering new ones; using tested theories and, when necessary, replacing them with better ones; and solving mathematical equations on which these laws and theories are based to unwrap the next mystery. Like children, we love asking questions, from the basic to the most sophisticated; we come up with outrageous ideas and then mercilessly scrutinize and discard most of them after applying logical rigor and observational tests. We are known for our lack of everyday practical skills combined with our love of logical deduction and synthesis.

At Maryland, I joined the Gravitational Theory and Cosmology Group, one of several faculty-led groups organized around specific areas of research. Groups include postdocs and students. (One group member was Charles Misner, a celebrated physicist who worked on the foundations of gravity; he was a former student of John Wheeler and had been a classmate of Hugh Everett at Princeton University, both of whom we will meet later in this book.)

During one seminar offered by this group, I heard a statement that shocked me. The speaker walked us through what the chances were that our universe would come into existence and concluded that the odds that our universe would form in the way that it did, with a big bang at high energies, were nearly zero! In fact, it was possible to calculate the odds, which the eminent British mathematician and theoretical physicist Roger Penrose (later a Nobel Prize laureate) had done in the late 1970s.

When Penrose calculated the likelihood of our universe spontaneously forming, he got a staggering number: 1 in $10^{10^{123}}$. Less than a one in a googolplex chance.

It was, to my mind, a completely ridiculous number.

Mathematicians joke that a googolplex is simply 1 followed by as many zeros as you can write before you get tired. The number is longer than the length of our entire universe.

If you are a cosmologist (and even if you aren't), there's something deeply distressing about Penrose's conclusion. Was the creation of our universe such a special event, produced out of such a unique set of circumstances, that it has never been and will never be repeated? Were we the virtually impossible winner of some bizarre cosmic lottery?

Along with Stephen Hawking, Penrose went even further. He and Hawking derived from first principles a logical argument in a theorem (a proposition that can be proved mathematically) that if our universe has been expanding since its creation, then it must have started from a point in space of literally infinite energy density—what is known as a singularity.

Hawking and Penrose's singularity theorem implied that scientists could never explore the actual moment of the universe's creation because nothing, absolutely nothing, existed *before* creation. That meant we could never replicate or identify the conditions that caused it, that the creation of our own universe was truly beyond our ability to study.

Which, of course, made it a thoroughly intriguing question for me.

When I lived in Maryland, one of my favorite weekend activities was to spend whole afternoons at a large bookstore in Bethesda browsing books on any subject from literature to philosophy to art—anything, that is, but physics.

Physics was reserved for weeknights and weekdays. On weekday evenings, I read everything I could from the scientific literature, and I even tried to reproduce the calculations. I wanted to understand how Penrose had arrived at what to me seemed like a preposterous conclusion about our universe; I wanted to try to follow the arguments that had convinced his fellow scientists to embrace the view that nothing existed before our universe did.

Although I was intrigued by Penrose's argument, I wasn't persuaded by his conclusions. I kept returning to Penrose's paper, dissecting and analyzing his reasoning, hoping either to be convinced or to find out where his reasoning might have taken a wrong turn. Certainly, I never thought that I would be able to solve the questions that Penrose and Hawking had walled off with their singularity theorem. I wasn't delusional. I was merely curious.

I quickly learned that Penrose's conclusion of a nearly zero chance of our universe coming into existence seemed rock solid. His finding was also deceptively simple. It was based on a fundamental law of nature—the second law of thermodynamics, which was predicated on the work of the esteemed nineteenth-century Austrian physicist Ludwig Boltzmann.

I had learned of Boltzmann's many contributions to thermodynamics and atomic theory during my undergraduate studies in Albania. (Incidentally, the professor who taught me thermodynamics went on to become Albania's president during the transition years.) However, it was not until later, when I started dismantling Penrose's derivation—the solution he obtained by solving those equations—that I fully understood, and appreciated, the significance of his work.

Boltzmann's discoveries were not simply a collection of equations bearing his name. They also formed a major stepping-stone in the development of modern physics: a breakthrough that revealed a crucially important relationship between the probability that an event would spontaneously come into existence and a concept known as entropy. Indeed, it was Boltzmann's probability insight that led to Penrose's ridiculous number—that the chances of our universe emerging randomly were nearly zero.

Simply put, Boltzmann's concept of entropy quantifies disorder. Imagine a kids' closet full of shirts in different sizes and colors. At a macroscopic level, the closet can be described perfectly well by its size, the color of its walls, and the number of shirts in it. One day, the parents decide on a rule for organizing the closet: all the shirts will be hung by size, from the smallest to the largest. Suppose the day after, the kids start hanging the shirts (or, more likely, throwing them on the closet floor) randomly. Unlike the parents' system, the specific ordering of which—arrangement by size—is unique, the children's system contains many different possibilities. But this information about the parents' or the children's arrangements is unaccounted for in the macroscopic description of the closet. In fact, every time the kids arrange the shirts differently and mess up the ordered closet, they create new configurations or, in physics jargon, new microstates. So, unlike the parents' unique and tidy system, a disordered closet has many microstates, since there are so many ways to disorder it. Despite the fact that the specific details of disorder are not captured in the macroscopic description of the closet, we can still deduce that overall, a disordered closet is not special because it is far more likely to be found randomly than a tidy closet.

The missing information contained in the collection of these microstates is Boltzmann's entropy. Entropy counts all the microstates that a system can possibly have without changing its macroscopic state, as in the closet example above. Entropy, therefore, measures what is hidden about a system. In the case of the closet, it minutely describes the fine details and disorder inside by means of a mathematical formula.

I already knew what Boltzmann's entropy was, but what I was really curious about was how Penrose connected it to the probability, or improbability, of our universe randomly coming into existence. How was entropy related to the birth of the universe? How did knowledge of the probability of the universe quantitatively emerge from knowledge of its entropy?

The answer is etched on the headstone of Boltzmann's grave in Vienna. At the very top, above a bust of the famous physicist, is an unusual epitaph—a mathematical formula:

$$S = k \, \text{Log} \, W$$

This is one of Boltzmann's most famous equations, what is known as his entropy formula. In this equation, S is the entropy of the system we would study; for example, the children's closet.

W is the number of microstates of this system; in our example, it is the number of all the possible ways of arranging the shirts inside the closet. Ln is the natural logarithm.* And *k* is a constant number, known as Boltzmann's constant, that makes the rest of the formula work. Simply put, entropy is proportional to (the *log* of) the number of microstates of a system. Or, equivalently, the number of available microstates of a system *W* is exponentially large with its entropy *S*.†

The formula on Boltzmann's headstone provides the first microscopic understanding of entropy in terms of the bits—the microstates—that make up a system. But until I revisited it in graduate school, I had overlooked the real meaning of his insight: that the number of these bits, the number of the possible microstates available in a system (as calculated by its entropy), is nothing less than a direct measure of the probability that this system will occur.

For example, there were many different ways, many microstates (*W*), for disorganizing a closet, but only a few ways of making it tidy and ordered. Therefore, if I were to randomly look at the closet, my chances of finding it in an ordered state would be very slim. The same principle applies to larger systems, up to and including the universe itself.

Any macroscopic system, whether a closet or the whole universe, has its own set of microstates through which it can occur

---

* Natural logarithm Ln is the reverse operation to the exponential. Ln is the logarithm to the base of the mathematical constant e ~ 2.7 such that $e^{(\ln x)}$ = *x*, for any value of the variable *x*.

† From the relation between the natural log Ln and the exponential, we can rewrite Boltzmann's formula as $W = e^{(S/k)}$. But the entropy of the early universe as estimated by Penrose is very low, therefore the probability *W* to start a universe like ours is as likely as 1 over the exponential of $10^{10^{123}}$—i.e., almost zero.

and be realized. If the universe at its creation moment has a large number of possible microstates through which it can come into existence, then the probability that it will randomly come into existence is high. Likewise, if the number of possible microstates through which a particular creation model can be realized is low, then the probability that it will occur is exponentially low.

Boltzmann's formula (which connects the entropy of a system to its probability of existence) implied that, in order for our universe to be exponentially less likely to come into existence by chance than any other universe we can imagine, as Penrose had calculated, our universe must have started from an exquisitely ordered state of very low entropy.

As I kept trying to connect the dots in my understanding of Penrose's derivation of the entropy of the universe, the story of our universe's unlikely existence became even more interesting. How can we even know what the entropy of our universe is? What information would calculating its entropy require? The entropy of the universe at each moment counts all the microstates of the universe, all the possible arrangements of its components. It reveals how disordered our universe is and what is not known about it.

Suppose that the closet in the previous example is as large as the universe. The shirts correspond to all the atoms and photons and stars and galaxies—all the matter, energy, and radiation in the universe. The amount of these components at present is inferred from astrophysical observations of our universe. Yet each time we exchange two photons from different sides of the universe, we have a new arrangement, a new universe microstate. Each time a supernova explodes and spews all its material into the universe, we have a new microstate, although the macroscopic universe as a whole remains the same. As in the closet example, if scientists know the universe's matter and energy content, they can count all the possible ways of spreading these bits around and calculate a system's entropy. This is what Penrose did. And as it turns out, the entropy of our present universe is not that large.

But there is always a catch, and in the case of our universe, it was this: The probability of our existence depended not on the entropy of the present universe but rather on the entropy at the moment of creation. This is tricky to discern, of course, since we cannot observe the moment of creation. Yet in order to estimate our universe's probability, Penrose had to find a way to deduce its entropy at its earliest moment. How did he achieve that? I had to find out.

In order to pinpoint the entropy of the universe at its creation, I needed to revisit what is probably the most important law of nature: the second law of thermodynamics.

The second law of thermodynamics states that the entropy of a system never decreases. Entropy always increases with time, no matter how much entropy you start with. In other words, the natural tendency for any system is to become more disordered, not less.

Imagine two adjacent rooms connected by an insulated door. Initially, the temperature in the first room is low (say, 40°F), and the temperature in the second room is high (say, 103°F). The second law of thermodynamics tells us what happens to the entropy of the rooms over time when the connecting door is opened.

As a result of air molecules moving between the two rooms, both rooms slowly reach the same average temperature. However, disorder grows, since we have more air molecules that have more space to roam around and create new arrangements. They can move from one room to the next and back. Heat is gradually transferred to the cold room, creating new microstates within both rooms. The more time passes, the more microstates are created in the two rooms, and the number keeps growing until the two rooms are the same temperature everywhere.

Furthermore, without external intervention, this process cannot be reversed. No matter how long we wait for the initially hot room to get hot again and the cold room to become cold again, it simply won't happen! The two rooms cannot spontaneously return to their original states. In other words, the entropy of the adjoining rooms keeps increasing over time—irreversibly.

This behavior is the essence of the second law of thermodynamics. Entropy growth over time is universal and irreversible, no matter the system. And the tendency of any system in nature is to try to reach equilibrium by increasing its entropy over time. The same conclusion is inevitable if I apply it to the whole universe: the entropy of the system—the whole universe—will increase irreversibly over time.

We know the entropy of the present universe, since, through our space- and ground-based astrophysical observations and a measurement of its expansion, we can count all of its content (mass, energy, and radiation), and with the help of the second law of thermodynamics, we can deduce that the entropy at the moment of creation must have been smaller than the entropy of the present universe. But precisely how much smaller was the entropy of the universe at the earliest moment relative to the present? Simply asserting that the state of the universe then had a smaller entropy than it does now does not provide sufficient information for estimating the probability.

There was another wrinkle. Penrose's argument for the ridiculously small probability of our universe's having come into existence, calculated from the formula on Boltzmann's headstone, depended solely on a number: the value for the entropy at the earliest moment in the life of the universe. Yet the second law of thermodynamics cannot provide an exact number for what the entropy was at the moment of creation. And this part of the puzzle became more complicated the further I delved into it.

Reconstructing the entropy of the universe at the moment of creation required tracking the cosmic evolution of the universe back in time, all the way to its inception. But therein lies the rub. The accepted narrative of the early universe given by the modern version of Big Bang theory (a story known as cosmic inflation), while wildly successful in explaining almost everything else about our universe, describes a universe of a very special origin, one of an exceptionally low-entropy state.

Cosmic inflation posits that in the blink of an eye, our tiny primordial universe, filled with high energy, became much bigger through a gargantuan explosion. It offers a compelling story for how a tiny universe can quite naturally grow large and later brim with life, and its exquisite agreement with observations helps explain why it remains widely accepted by scientists and the public to this day.

But Penrose's paper raised lots of doubts about the theory of cosmic inflation. Our universe's high-energy but very low-entropy initial state posited by the theory of cosmic inflation implies that the probability of our universe beginning in this manner was as small as it could possibly be. By pointing out this wrinkle in the cosmic-inflation theory, Penrose's argument posed the greatest threat to the validity of cosmic inflation being the progenitor state of our universe.

This, in a nutshell, was the infamous problem of the origin of our universe.

My simple plan was to zoom in on the details of cosmic inflation and the process of what happened afterward. I wanted to familiarize myself with previous attempts at solving this problem and, especially, understand why and where those attempts failed. If the issue of our special origin was an artifact of cosmic inflation, then perhaps we needed to discard that theory and replace it with a better model of creation. As I wondered at the time, did the apparent unlikelihood of our universe coming into existence indicate a different problem, even a fundamental problem, with the generally accepted explanation of our origins? Or were we entirely missing the point and asking the wrong question? It turned out the answer to both was yes.

2

How Did Our Universe Start?

T HE TRAJECTORY OF my own life was changed by a momentous historical bang: the fall of the Berlin Wall in 1989. Although Albania was not a part of the Soviet Union, the tide of transformation was impossible to ignore. Students began rallying for free speech and pluralism in 1990, but the real revolution arrived in 1991. That February, a group of university students initiated a hunger strike in their dormitories, demanding that the current ruler, Ramiz Alia, give up power.

At the time, I was an undergraduate at the University of Tirana, but I lived at home with my parents, so I didn't know about the plans for the hunger strike until the day it happened. As it dragged on, I sat with my friends and their parents outside the dorms to keep the strikers company.

When the students' medical conditions deteriorated, a few hundred miners walked fifty miles to Tirana, determined to free the students and overthrow the government. My family and I joined the students, parents, and miners as they marched from the dorms to the center of the city to protest. Along the way, the crowd kept swelling as many other citizens joined the demonstration. Young soldiers, only eighteen to twenty years old, looked confused as they hesitantly pointed their guns at the crowd. A touching moment that day was when the numerous elderly Albanians who were marching courageously approached and hugged the soldiers. They pressed their chests against the gun barrels, saying: "You are our blood, our kids, we are fighting for your future too. Your own family must be somewhere in the crowd. Don't shoot at them. Drop your weapons and come and join us."

By the time we reached Tirana's central square, the crowd had expanded to many thousands. The army was deployed to close off the streets and stop people from joining the protests. With the streets shut, we were trapped in the square. Helicopters flew overhead, and snipers took up positions on the rooftops. People began pulling up cobblestones and ripping marble steps from stairways inside buildings to be able to defend themselves in case the army shot at them. Had the soldiers opened fire on the crowd, it would have been a massacre. But sensing that freedom was only hours away, after a fifty-year wait under the worst dictatorship in Europe, the protesters did not turn back.

At this tense moment, my younger brother disappeared into the teeming throng. My dad and I sent my mom home to wait for my brother, but as it turned out, she was wasting her time; he had taken up a position at the front line of the protesters. All around us, the crowd was chanting, "Freedom! Down with Alia. We want Albania to be a democracy; we want Albania to be like the rest of Europe," over and over again.

Police, soldiers, and special security forces in long coats leading fierce Alsatian dogs swarmed the square, but the protesters kept chanting. Then the demonstrators began pulling down the giant bronze statue of Albania's first Communist ruler, Enver Hoxha, which had stood for what seemed like forever in the central square.

The soldiers were poised for action, expecting to receive the order to shoot. But for some reason, as the colossus toppled, their radio communication was cut off.

Over the din, we could hear the soldiers screaming at one another, asking what the order was and why their radios had suddenly gone silent. Later, a rumor circulated that Ramiz Alia had cut the signal to keep the generals from taking matters into their own hands and giving the order to fire without his permission. Incredibly, that day, both the protest and Communism in Albania ended without a massacre. My brother later brought home chunks of marble from the pedestal of the statue of Hoxha, a small reminder of the day that Albanian democracy almost died in the womb.

The months between the first student strikes and Albania finally shedding Communism had been brutal and chaotic. Thousands of people jumped over foreign embassy walls to seek protection, including all the students at the University of Tirana—except me and one other classmate. (It is estimated that between 170,000 and 300,000 people left the country during that period; the only two who failed to get out were the ones who jumped the walls at the Cuban embassy—the guards there handed them back to the Albanian authorities.)

On the night that my classmates had picked to leave, we all gathered in Tirana's main park at sunset to say our goodbyes. Until then, we had shared every little thing we had: pocket money, lunches, pencils, and notebooks. I tried to stop them from leaving by suggesting that they should finish their degrees (we still had a year and a half to go) and explaining that where they were headed was not like the *Dynasty* series we had recently been secretly watching on TV using homemade antennas. I argued that Communism was over and we had nothing to fear. They, in turn, pleaded with me to go with them, teasing me that I

thought too much about everything. But I had made my decision. I couldn't abandon my parents, and I wouldn't be a beggar
at the mercy of strangers for food and shelter. I had (clandes-
tinely) read enough English books at the private library of the
League of Writers and Artists, my mom's workplace, to know
that life was hard in the West too. After we'd talked for hours,
they asked me and the other classmate who was staying behind
to find their parents and tell them not to worry.

Around midnight, I walked with my friends to the German
and French embassies and wished them luck, then watched them
climb over the fences and disappear. The embassies would
arrange to take them by plane or ship away from Albania.

Night after night, Tirana was alive with young people planning
their escape and parents who had come from all over the country
searching frantically for them in the dark, first in the center of the
city, then in the gardens along Embassy Road. The city sounded
as if it were in mourning, filled with rushing footsteps, whispering
silhouettes, and crying and sobbing.

I remember one man approaching me as I headed home one
night. He was weeping quietly and cradling a pillow next to his
face. He told me he had heard a rumor that his son wanted to
leave, and he had traveled for four hours to reach Tirana to try
to stop him. He had gone to his son's dormitory and found his
bed untouched. He described his son and asked if I had seen
him; I hadn't. He sniffed the pillow, showed it to me, and said,
"This is all I have left of my son—this has his smell."

When I left for the United States two years later, I carried the
marble pedestal fragments that my brother had given me as a
reminder of my new freedom and the lack of it in my previous
life. I still keep them, and not merely as mementos. The lessons
they contain about intellectual courage and the need to confront
orthodoxy remain as relevant to me today as when I left Albania.

And during my graduate studies at the University of Maryland, these lessons were foremost in my mind as I dug deeper and deeper into the questions unleashed by Penrose's paper.

The more that I learned about my new field of study, the more uncertain I became about the validity of its dominant arguments concerning the big questions about our universe. The biggest question was, of course, how our universe was created and what had been there before. And the prevailing answer to this question did not satisfy me at all.

In 1997, after I earned my master's degree at the University of Maryland, I began studying for my PhD at the University of Wisconsin–Milwaukee. I wanted to focus on the quantum aspects of the early universe, and I chose UWM because it has one of the strongest theoretical physics groups, especially in quantum physics, in the United States. In particular, I wanted to work with Leonard Parker, a world-renowned theoretical physicist and one of the founders of a new discipline called quantum field theory in curved space-time.* Parker's work demonstrates that as the universe expands and changes curvature (shape), the corresponding change in its gravitational field is converted into an energy that produces particles that populate the universe. This field is one of the most groundbreaking areas of scientific study today.

Professor Parker was among the kindest and most modest people I ever met. He welcomed me as his student, and he and his wife treated me as if I were one of their own children. I spent

---

* The curved space-time unites the three-dimensional space (length, width, and height) with the dimension of time into a unified object, the four-dimensional space-time—the shape of which, as we will see, is curved and corresponds to a gravitational field according to Einstein's theory of relativity. Quantum theory is something you will learn about in chapter 3.

three graduate-school years in a cold, windy, and snowy city with beautiful architecture influenced by Frank Lloyd Wright, but I never noticed the cold because of the warmth of the physics group and the people of Milwaukee.

During my years in Milwaukee, I came to understand more completely why cosmic inflation, despite Penrose's argument that it had a near zero chance of igniting a universe, had nevertheless earned its central place as the theory of the universe, commonly known as the standard model of cosmology. In fact, in my doctoral dissertation, I investigated the alternative theories to cosmic inflation (the gargantuan explosion of an infant universe filled with high energies to produce a big universe like ours) that had been proposed by the theory's opponents. Those opponents included Hawking and Penrose. The investigations of scientists who came up with alternative models for the creation of our universe were very important in scrutinizing cosmic inflation. Such alternative models included tunneling of an infant universe through a gravitational field and a group of scenarios of possible phase transitions that could have potentially produced a universe like ours.*

The more I scrutinized cosmic inflation's foundations, the more convinced I became that despite the problem of the unlikeliness of our universe coming into existence, the theory still offered by far the most logical and elegant explanation for the fundamental properties of our universe. This is in part because it preserves the integrity of major theories such as Einstein's theory of general

---

* An infant universe's phase transitions might be similar to the transitions in phases of matter—solid, liquid, gas, and plasma—that we are familiar with in daily life. For example, much like water goes from liquid to gas through boiling or from liquid to solid ice through freezing, an infant universe might change from one state to another depending on the forces acting on it.

relativity and the assortment of models collectively known as Big Bang theory—theories that otherwise struggle to account for how to obtain our present universe from the strange conditions that existed at the dawn of the universe as we have come to understand them.

Cosmic inflation uses Einstein's theory of general relativity to link the universe's matter and energy to its curvature—that is, its shape—and its expansion. Einstein produced this theory in 1915 while standing on the shoulders of mathematical giants.* He came to regret the name *general relativity* because it seemed to contradict his belief that the world existed independently of human observation. According to Einstein, what happens in the universe is definitely not relative to who is observing it or how the observer is moving; reality, he felt, must be objective.

To make his theory hang together, Einstein relied on two postulates. The first one is that the speed of light is the absolute limit of speed that any object in the universe can travel. With his second postulate, Einstein united three-dimensional space (height, width, and length) and one-dimensional time into a single entity, which he called space-time. Our universe, Einstein asserted, exists in a four-dimensional space-time.

Einstein's second postulate is the amazing insight that we can essentially trade the force of gravity for the shape of space. The way that Einstein achieved this trade-off in his theory of general relativity is beautifully simple: According to him, the gravitational

---

* For his theory of general relativity, Einstein relied heavily and used the foundational work on curved space-time geometries discovered by Bernhard Riemann in Germany; for his postulate of space and time being on equal footing, he relied on the work of Hermann Minkowski also in Germany and Nikolai Lobachevsky in Russia.

force of matter and energy in the universe tells space-time how to curve, and curved space forces objects and light to move along certain paths that follow the curvature of space. His theory replaces the gravitational force exerted on any object produced by all the matter and energy in the universe with the curvature of space-time shaping the paths along which that object moves. Gravity is curvature.

Einstein's profound insight is not hard to visualize. Imagine you have set up a perfectly flat hammock in your garden. The fabric of the hammock is the space-time in this example. Now, if a person sits or lies down on the hammock, it will bulge downward—that is, it will curve according to the person's position and weight. In our metaphor, the person is the matter-energy content, and his or her body weight and size determine how the shape of the hammock—the space-time—curves. Thus, as the physicist's adage goes, "matter tells space how to curve." Importantly, the shape of the hammock—the curvature of space-time—indicates how much matter-energy it contains. If you sat below the hammock, you would be able to judge the size and weight of the person in the hammock above simply by evaluating the curvature of the fabric.

If the hammock represents the curvature of space-time of the whole universe, and the person's weight on it represents the energy of the universe at the time of cosmic inflation, then, according to Einstein's theory, the energy of inflation determines how fast the universe can expand and what shape it will take. In this way, Einstein's general relativity laid the foundations for the theory of cosmic inflation, which would in turn use Einstein's general relativity to explain the strange circumstances that existed at the dawn of our universe.

• • •

Since Einstein's day, scientists have taken precise measurements of the shape of the universe and all that's inside it; this has allowed us to reconstruct its history at various epochs all the way back to its birth. Using Einstein's equations, we can reverse-engineer what the universe looked like and how fast it expanded at every moment in its past. In the far past, the universe became microscopic as it approached the creation moment. Alas, at that point, Einstein's equations break down.

This is the major downside of Einstein's theory of relativity: it becomes invalid under conditions of very high energy densities—the type of energy density that one might find at the center of a black hole, for instance, or that existed at the first instant in the life of our universe. In fact, using Einstein's equations to find the shape of the universe as a function of the energy contained within it at the universe's earliest moment, just before cosmic inflation flicks on, generates a disappointing answer: it predicts that the universe started off as a pinpoint, a singularity that pinches off the very fabric of space-time.

This breakdown of Einstein's equations at the birth of the universe is known as the Hawking-Penrose singularity, after the two iconic physicists who generated the theorem (a scientific milestone alluded to in chapter 1). Time stops at this singularity—there is no "before"; clocks freeze. Space stops there—there is no beyond. According to Hawking and Penrose, nature forbids scientists to explore the moment of creation, let alone look past it, because nothing, absolutely nothing, existed before creation.

There is a century-long history of surprises that scientists came across whenever they tried to obtain solutions to Einstein's equations at the earliest moments in the life of our universe. In 1922, Alexander Friedmann, a Russian theoretical physicist and mathematician, used Einstein's equations to demonstrate that they produced an expanding rather than a static, unchanging universe. He wrote to Einstein sharing his calculations, but Einstein was unconvinced.

Five years later, Georges Lemaître, a Belgian astronomer and Catholic priest, independently proposed the first model of an "exploding" universe that started small and grew exponentially. Observing that galaxies were moving away from one another, he conjectured that our universe must have been born from a "cosmic egg." Lemaître's finding was verified by Edwin Hubble (of subsequent telescope fame) two years later; he also proved that galaxies outside of our Milky Way were constantly moving away, or receding, from one another.

In the 1940s, Lemaître's version of the exploding universe was developed into a model by George Gamow, a Russian-born nuclear physicist and a celebrated author of popular science books who had defected from the USSR and eventually ended up at the University of Colorado Boulder. (Gamow's doctoral adviser, not coincidentally, had been Alexander Friedmann.) Like Lemaître and Friedmann, Gamow used Einstein's theory of general relativity to link the matter-energy content of the universe to its curvature and expansion. The result was something that we now know as the "Hot Big Bang."

Gamow's theory of the universe's creation was both elegant and attention-grabbing. Relying on Einstein's equations, he envisioned the universe in its infancy as a tiny vessel, roughly the size of an atom, filled with a "hot primordial soup of radiation" that "banged" into being and grew large over time. He went so far as to predict the existence of radiation relics in our sky left over from the time of this Hot Big Bang.

Gamow's Hot Big Bang marked the emergence of a new field in physics: cosmology. But his model and the other Big Bang models that followed, all of which depended on hot radiation to make the universe expand, had severe shortcomings. Specifically, these models failed to explain three crucial features of our universe: its flatness, its homogeneity, and the uniform distribution of all the matter in it. This is the failure that cosmic inflation was intended to compensate for.

If you look at the sky through a telescope lens, you will see the same distribution of matter and light no matter where you swivel the telescope. But this phenomenon isn't new; it's existed for the universe's entire history. I like to picture our skies at each moment in time as a canvas that has been sprayed randomly with paint composed of light rays and particles. The distribution of the paint blobs and the blank areas of the canvas is more or less the same everywhere, on the left, on the right, in the center, at the top, and at the bottom.

Yet no matter how many tweaks were introduced to Gamow's Hot Big Bang theory, it could not generate the same uniform and homogeneous universe that we see in our skies. The primary reason for this discrepancy was the type of energy that Gamow had chosen to bring his universe into existence. A primordial universe filled with hot radiation could not grow fast enough to reach the size of our present-day, very large universe. The only way Gamow's Hot Big Bang could work was if it started with a huge (in relative terms) primordial universe, one roughly the size of a helium atom.

This initial size is what presented the insurmountable problem that, decades later, cosmic inflation so elegantly solved. It is one of the reasons why, despite Penrose's argument on the unlikeliness of our universe coming into existence, cosmic inflation continues to be the foundation of cosmology.

In Gamow's time, the trouble was that the hypothetical primordial universe was too large. During the brief instant of the Hot Big Bang explosion, light and particles would not have sufficient time to traverse the helium-atom-size universe. They would be able to travel only a short distance, a minuscule fraction of this universe, and that means that they could never become uniform or homogeneous.

Imagine a scenario where the primordial universe is comparable to the size of the United States (3.79 million square miles). In this universe, the maximum travel speed is ten miles per hour, and the Hot Big Bang lasts for an hour. During that hour, light and particles can travel only ten miles from their location to connect to other objects. Anything beyond ten miles is outside their reach. This means that, in our scenario, New York and California would not be able to exchange any information, nor would they know of each other's existence. In this example, millions of different regions in the United States would be completely disconnected and would evolve independently of one another. The end result would be that, when we looked up at the sky, rather than seeing one large uniform universe, we would see a mosaic made up of many, completely independent skies, each with potentially radically different distributions of stars and planets. The many disconnected regions of the Hot Big Bang primordial universe would be entirely independent of one another, and so would their temperatures. Fast-forwarding to the present day, scientists would expect our skies to be a patchwork of different temperatures and matter distributions, literally trillions of different skies in one. But that is clearly not the case with our universe; everything that we can see is arranged uniformly and homogeneously.

As it turns out, the initial size of the universe required by Gamow's Hot Big Bang theory is only part of the problem with that model of creation. By the 1960s, for reasons I will explain in a later chapter, scientists knew that the shape of space in our universe is flat. This discovery relied on basic geometry and the fact that whenever we look far away at distant stars, we are looking back in time—we are looking at the moment when the star emitted its light, and that might have been billions of years ago.

To understand how something three-dimensional, like our universe, can be described as flat, try this simple thought experiment. Pick three bright stars and use them to draw an imaginary triangle in the sky. (This triangle will actually be three-dimensional, because, although our sky may look two-dimensional from the ground, it is not; each star is located not only in space but also in space-time, so the starlight is, in fact, an incredibly complex layering of time and space, converging in the twinkling lights that we can see with the naked eye.) Now consider the angles in the triangle you've created. If our universe were spatially curved like a sphere, then the angles in your triangle would add up to more than 180 degrees, as shown in the top panel of figure 1. This is an example of a "closed" universe. If our universe were curved in space like a saddle, it would be an "open" universe, and the angles of a triangle drawn on that space would add up to less than 180 degrees, as in the bottom panel of figure 1. But our universe is neither open nor closed. We live in a spatially flat universe (with zero curvature) where the angles of a triangle drawn on its sky add up to exactly 180 degrees, as in the middle panel.

The revelation of a flat universe posed a thorny problem for Gamow's elegant theory of creation. If the universe had started with a Hot Big Bang, obtaining a flat universe out of it using Einstein's equations was hard. It required a lot of unjustified complications and contrivances that were not naturally motivated by basic physics. Such a model would not be considered plausible or attractive.

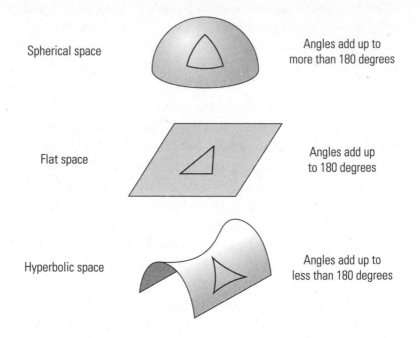

Spherical space — Angles add up to more than 180 degrees

Flat space — Angles add up to 180 degrees

Hyperbolic space — Angles add up to less than 180 degrees

Figure 1. From top to bottom, a spatially closed, flat, and open universe.

Given what was at stake, the failure of Big Bang models to explain the three key features of our universe—its flatness, homogeneity, and uniformity—was a big deal. Was the Hot Big Bang wrong? Gamow's theory seemed to be imploding—until a new theory swooped in to save it.

In the late 1970s and early 1980s, two young scientists, Alan Guth, then a postdoctoral fellow at Cornell University, and Andrei Linde, working behind the Iron Curtain in Moscow, independently came up with a clever idea to fix the problems that dogged Gamow's Hot Big Bang theory. Their solution was dubbed the "inflationary universe," or "cosmic inflation" for short, and it is considered a masterpiece of twentieth-century physics.

Guth and Linde's cosmic inflation solves the original Hot Big Bang problems very simply: it replaces the primordial hot soup of radiation that Gamow thought had banged the universe into creation with a hot soup of energy. Specifically, Guth and Linde postulated the existence of a slowly rolling primordial particle at the Big Bang moment that they dubbed the "inflaton."* Starting a universe with a hot soup of energy from the slowly rolling inflaton seems to make all the problems vanish.

Central to the power of Guth and Linde's theory of cosmic inflation is the nature of the inflaton energy that drives the expansion of the universe. It is a special kind of energy, sometimes referred to as "vacuum energy." Guth and Linde's solution to all the problems that the Hot Big Bang models faced relies on a crucial property of vacuum energy: negative pressure, a repulsive gravitational force that tends to blow things up—that is, inflate them. Thus, a primordial universe filled with vacuum energy explodes and grows much faster than one filled with the regular radiation posited in the old Big Bang models.

In fact, according to cosmic inflation, our infant universe grew so quickly that within an instant—$10^{(-45)}$ seconds, to be exact—the size of the original universe, whatever it was, increased by about twenty orders of magnitude. That is, its original size multiplied by 10 with 20 zeros. To get an idea of the scale involved, first imagine the very thin wall of a soap bubble (only a few nanometers thick). Next, picture the distance from the Earth to the sun (roughly 146 million kilometers, or twenty orders of magnitude bigger than the thickness of our soap bubble). With

---

* A slow-rolling inflaton is analogous to a golf ball that is hit onto a sandy beach and moves ever so slowly, maintaining roughly the same energy.

cosmic inflation, the wall of the soap bubble expands the distance
of the Earth to the sun in a minuscule fraction of a second.

So, unlike the original Big Bang models in which the primordial
universe had to be, in our analogy, the size of the United States in
order to evolve billions of years later into the presently observed
size of our large universe, a cosmic-inflation universe needed to
be only the size of Manhattan to achieve the same expansion. It
blows up so quickly that, within a fraction of a second, Manhattan
becomes as large as the entire United States.

All the stuff contained inside this primordial universe, whether
waves of light or particles, stretches with the universe's accelerated growth. And during inflation, these elements remain in
communication with one another across the span of the universe,
continually equalizing their temperature and spreading homogeneously everywhere inside. Particles being able to maintain
communications throughout the universe implies that the chain
of cause and effect, that sacred principle of nature known as
causality, is preserved.

Next, as the universe inflates to trillions of times beyond its
infant size, it stretches itself to flatness.

Thus, in contrast to Hot Big Bang models, the cosmic-
inflation model achieved flatness, homogeneity, and uniformity
all in one, in the most natural way.

Guth and Linde's explanation for cosmic inflation relied heavily
on a new area of physics that arose in the twentieth century,
quantum theory, which I will write more about in the next chapter.
Both the energy field they postulated for cosmic inflation and
the microscopic universe that existed in the first moments are
quantum in nature. It turns out that the quantum energy of
cosmic inflation that started the universe also has an extremely
low entropy, which, according to Boltzmann's formula—as
Penrose pointed out—implies a very small probability of existence.

Therefore, the very conditions that they had declared were present at the creation of the universe were the same ones that made the universe's creation incredibly unlikely.

As this connection between the energy of cosmic inflation and the fact that inflationary energies will always have a very low entropy clicked, it triggered a new picture in my mind.

Cosmic inflation is amazing, it's brilliant, it is probably the finest origin story of our universe so far—of that I have no doubt. It offers the most logical explanation for the fundamental properties of our universe in what seems to be the most natural way; astrophysical observations agree exquisitely well with its predictions. It is correct.

But my personal belief is that, although correct, cosmic inflation is an incomplete theory. It requires us to accept an impossibly unnatural assumption: that our universe began in the most special way possible, with a perfect inflaton in a perfect hot soup of energy in a smooth space that was the smallest possible size it could be without Einstein's theory of gravity breaking down (something physicists call a Planck length). So, while cosmic inflation, based on one assumption (the existence of an inflaton energy in the first instant of the life of the universe), provides the perfect explanation of how a tiny universe evolved to its present state, the entire story hung on one mystery: What gave that inflaton the energy that jump-started the inflationary process?

I was still only a graduate student gathering information step by step. But here, sitting before me, was an increasingly intriguing mystery—one that I found hard to resist.

3

A Quantum Leap

S OCIALLY, I HAVE never been a daredevil, but academically, I probably qualify as one. And I am fortunate that, long before I arrived in America, my professors—in their own way—encouraged this quality in me.

One of the best and most feared math professors at the University of Tirana was Professor Bardulla. His family, we heard, had been killed by the Communist regime. He was harsh and short-tempered, and he never smiled; he wore an old suit covered in cigarette burns, and he was almost always drunk, even during lectures. He often brought a bottle with him to class. But even drunk, he was still sharper and more quick-witted than many of his colleagues. It has been quite a long time, but I still remember his permanently red eyes sparkling with a mix of powerful intellect and bottled-up pain.

I took Professor Bardulla's final exam and finished it earlier than the other students. When I stood up to leave, he looked at me, furrowed his eyebrows in disgust, and quietly asked, "What do you think you are doing?" I told him I had tried a different solution than the one he had taught, and with this shortcut, I needed less time. He took my papers and read them line by line.

My heart pounded as I watched him turn the pages and wince. He reached the end, put the exam away, and said nothing. He just stared at the floor while I waited to be scolded. Then he looked at me sternly and said, "I need to give you a piece of advice that you must remember for the rest of your life. Never, ever do this again. Never try a new method or a new solution in an exam. It is too risky. If you get it wrong, you fail. Do you understand? At least with the known solutions, you can get partial credit."

In a weak voice, I asked, "Was my solution wrong?"

"No," he said. "It is correct, and I like it."

I like it. Those three words encouraged me to set off on the path of intellectual risk-taking. (And although I couldn't have anticipated my future, I'm very glad I succumbed to my attraction to math. It gave me the confidence not to shy away from a physics problem when the math appeared too complicated.) It helped, too, that the physics and math professors in Tirana took pride in their students and fed our curiosity in any manner that they could. This attitude was not unusual—in Communist Albania, knowledge was a way of protesting and rising above the suffocating regime. Albanian intellectual society both deeply respected knowledge and was thirsty for it, especially since there were no other distractions or entertainment available.

In many ways, the regime's iron fist made intellectual ideas more alluring, not less. For example, forbidding Western literature didn't make us less interested in it; it had the opposite effect—it increased our curiosity and desire to read and learn. Highly skilled people in different professions found underground ways to escape boredom and censorship; groups of them regularly met for coffee to share advances and discoveries in their different fields. My parents' circle of friends ranged from medical doctors to scientists, writers, composers, and artists. I loved listening to their conversations; they broadened my horizons and encouraged me to respect and nurture interests in other areas.

Perhaps the combination of my upbringing and the deprivations of the regime sparked my habit of drilling down on questions that intrigued me to the exclusion of almost everything else. But whatever its origins, this quality came to define my graduate studies—and sent my research in a different direction, away from what some of the leading minds in my chosen field considered mainstream.

Scientists study physics to understand how things work and establish the principles and laws of nature that describe the operation of our world. Indeed, classical physics, which explains the visible, macroscopic world, predicts outcomes with 100 percent certainty. In the parlance of physics, it describes a deterministic world.

But gradually, as the twentieth century unfolded, it became clear that certain phenomena in the microscopic realm could not be explained by the laws of classical physics. Another set of principles operated in that domain, the chief characteristic of which was not determinism but almost maddening uncertainty. Over several decades, an entire branch of physics arose to deal with this uncertainty: quantum theory, whose mathematical laws and operations are described by quantum mechanics. And as I delved deeper into my PhD studies in Milwaukee, I began to suspect that the answer to the universe's origin might be hiding somewhere in the realm of quantum mechanics.

As I've written, the trajectory of my life had been changed by events beyond my control. Had any of the events that brought me to the present been different, my life would have taken a different path.

Today, whenever I am teaching quantum theory to my students, I cannot help but think that my own life bears some resemblance to a quantum reality, a collection of chances and events, each of which, had they turned out differently, would have taken me down a very different path. Had the Berlin Wall not fallen, I would probably be living in a dictatorship, probably forced into internal exile like my father. Had I submitted to the peer pressure of jumping the embassy walls with my friends, I would not now be researching the universe; I would probably be living somewhere in Europe, perhaps not even having finished college. Had I not submitted a Fulbright application, I might never have left Albania. Had I not accepted an assistant professorship at the University of North Carolina at Chapel Hill just over a decade after I arrived in the United States and four years after I completed my doctoral dissertation, I would most likely be living in some other state or some other country. And had I been more "practical" in my choice of research problems and not given in to my curiosity about the creation of the universe, the only theorizing I would be doing about early cosmology would be over coffee or cocktails. Had any of these events been different, my life would have been different. And that is the essence of the quantum world, the world from which our universe was born.

As if the options and uncertainties at the level of one individual like me were not confusing enough, the discoveries that became quantum theory—arguably the most profound theory in the history of science—combine to describe the whole world in terms of a staggering, multilevel, almost incalculable number of uncertainties, a mind-boggling concept that has driven our greatest scientists to the edge of reason. Consider this one example: In the quantum world, it is possible for a single object to exist in two different states—to be both a particle and a wave and perpetually fluctuate between the two. Furthermore, the entire quantum world is based on probabilities—the chances of having different outcomes to the same questions. These qualities of the quantum world defy reason, yet as far as physicists are concerned, they are a fact of life, the same as gravity or the changing of the seasons.

Most of the discoveries of quantum theory emerged in the twentieth century. In the twenty-first century, quantum principles underlie every aspect of the groundbreaking discussion of the first and last moments in the life of the universe. This revolution in thought is all the more remarkable when you recall that at the end of the nineteenth century, many physicists did not even believe in the existence of the atom.

Most important for our purposes, the breathtaking rise of quantum theory laid the groundwork for subsequent research into the tiniest scales in the universe, including the microscopic origin story of the universe itself. Understanding how this field of physics originated and what implications it holds is key to comprehending why the mystery of the universe's origins unraveled as it did.

Much of the credit for the origins of quantum mechanics goes to the German scientist Max Planck. My dad and I derived great pleasure from listening to radio broadcasts of Bach in the dark of night, and like us, Planck was a lover of music. But ultimately he pursued physics, despite being told that there was almost nothing new to be discovered. In the late nineteenth century, no one would have suspected, not even Planck himself, that he would become a disrupter who threatened centuries of classical physics and a revolutionary who rang in the new century, the belle époque in physics, with a new theory of nature called quantum mechanics.

Classical physics describes a deterministic world. But during Planck's time, it became clear that certain phenomena in the microscopic realm could not be explained by the laws of classical physics. Another set of principles, those of quantum theory, operated in that domain.

Conservative in his thinking and renowned for his scientific integrity, Planck started off as a strong proponent of classical physics. Like his protégé Einstein, he admired James Clerk Maxwell's classical theory of electromagnetism, which unified electricity and magnetism and is considered one of the great breakthroughs of the nineteenth century. Maxwell's theory describes a continuous stream (spectrum) of energies from both undulating electric and magnetic fields. These radiation energies are often referred to interchangeably as "light waves." Light waves would prove to be at the center of Planck's conversion story—and of Einstein's.

Light waves share a series of common properties with all the other types of waves (shown in figure 2). Whether they are made of light, sound, or seawater, waves have three common features: wavelength, which is the distance from one crest of a wave to the next; frequency, which measures how many wavelengths (or crests) pass through a fixed point each second; and amplitude, the strength of the wave, which is measured by the height of the wave's crest. But unlike waves that require a medium—a material in which the wave can travel and be sustained—Maxwell's electromagnetic waves can propagate in empty space-time, in a vacuum. They can sustain themselves by transforming from one form to another as they travel, an electric field becoming a magnetic field and vice versa, in cycles. (The only other waves known in nature that can spread through a vacuum are gravitational waves—a type of wave that, in Planck's day, had not yet been discovered.)

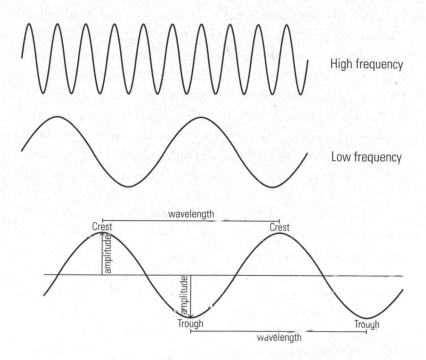

High frequency

Low frequency

Figure 2. Characteristics of a wave defined by their amplitude, frequency, and wavelength, as shown.

At first, their admiration of Maxwell's work prompted both Planck and Einstein to resist the very theory they were helping to establish. At the same time, they shared a combination of traits exhibited by all great scientists, then and now: the courage to advance radical, groundbreaking ideas by deploying rigorous skepticism, a skepticism that requires the merciless and arduous scrutiny of every detail of their ideas. A good scientist is both a rebel and a conservative, a creator and an auditor, all in one.

Planck ultimately broke with classical physics due to the influence of Boltzmann's atomic theory and Boltzmann's probability relation to entropy, a theory that he had opposed and resisted for a decade. Planck was forty-two years old when he declared, in October 1900, that light had a "double personality." It was not only a radiation wave, as Maxwell envisioned it, but also a collection of photon particles (so named by Einstein, from the Greek word for "light"). This insight would turn physics on its head and pave the way for some of the most revolutionary research about the cosmos.

Planck published the hypothesis that light was a collection of discrete quanta (in layman's terms, bundles of radiation waves) that resemble particles.*

Planck's collection of energy quanta was the first-ever description of a quantum particle. His insight revealed a new component in the DNA of nature: the wave-particle duality of quantum mechanics. In time, Planck's quanta would form the basis for a new field of physics—and a host of new theories and discoveries about the origins of the universe.

Planck also replaced Maxwell's continuous spectrum of waves with a collection of steplike (quantized) levels of energy. According to him, energies of light waves do not roll continuously into one another but instead jump from one level to the next in discrete steps, each step being one finite quanta at a time.

To picture Planck's quantized spectrum, suppose you are that light wave and the energy levels are the floors in a building. Imagine you are going from the second floor (high energy) to the first floor (lower energy). You can do this by taking the elevator, which will

---

* Each quanta or light particle with frequency v carries a unit of energy $E = hv$. In Planck's expression, h is a numerical factor, a universal constant of nature, that he introduced "by hand." It is now known as the Planck constant.

continuously lose height in smooth incremental bits, or by taking the stairs, in which case you are going down in discrete finite steps, one step at a time. You cannot take half a step or a quarter of a step. If you don't want to fall, you have to take exactly one step—one quanta. In our analogy, a building designed by Maxwell would not have a staircase; a building designed by Planck would not have an elevator.

Planck's contribution was brave and significant. It laid the first block in the foundations of quantum theory out of which a new theory of nature would emerge. It also proved crucial to my investigation in decoding the origin of the universe. Because Planck's work revealed that at its earliest moments, our universe was not just an object—it was also a wave.

The next step in the development of quantum mechanics was to prove that wave-particle duality existed beyond light. Here, two seminal contributions stand out. The Danish physicist Niels Bohr pioneered an atomic model in which electron particles circled the nucleus of the atom in well-defined orbits. Meanwhile, the French physicist Louis de Broglie hypothesized that electrons could behave like waves as well as particles.

De Broglie's conceptual leap led us to view all the particles and light in the universe as, intrinsically and simultaneously, both waves and particles. All particles, including you and me. Including the whole universe! We are simultaneously stardust and starlight. We are all waves!

I'm making light (so to speak) of a profound statement about the implications of the wave-particle duality of matter and light, a universal property of the universe that immediately begs some simple questions: If you and I are waves, how come we don't see a wave trailing us as we walk down the street? Why can't we glow like the stars do? If one's dual self, the quantum shadow, can be a wave, why can't a person travel through glass and walls like light and sound waves do?

If you are entertaining the idea of testing this last question at home (as I have inadvertently done myself in my more absent-minded moments), I can offer a scientific reason why you shouldn't: your wavelength is way too small to pass through the glass or the wall. A human of average weight walking at normal human speed would have a wavelength of about 10^{-36} meters (this is 1 divided by 10 followed by 36 zeros), sufficiently minuscule to be beyond observation, much smaller than the thickness of any wall you wish to penetrate, and thus irrelevant to your own experience. Any object bigger than a human, such as a planet, a star, or a galaxy, would have an even smaller associated wavelength, so we can safely ignore their wavy nature. Big, heavy objects have very short wavelengths. Light, tiny objects have long wavelengths. Consequently, in classical physics, which governs the realm of macroscopic objects, objects are just objects and waves are just waves, and neither can be simultaneously both a wave and an object.

I learned the lesson that we humans cannot easily switch into waveforms on the night before the defense of my doctoral dissertation in Milwaukee. Physicists are often lost inside their own heads, and certainly I was on this particular evening. I had gone to my neighborhood bookstore, which had a coffee shop, and sat there for hours rechecking my formulas and explanations. I finally left, exhausted, but I was still lost in thought, turning pages of equations in my head for a final check as I walked. So, in a manner of speaking, I existed in my head as only a waveform but neglected to take into account the fact that I was also a physical object. I reached the main street that led to my building and began to cross, but, as was my habit, I never looked up to see if the pedestrian light was red. (It was.) Worse, but typical of me, as I stepped onto the sidewalk, I bumped into someone. I apologized to him without looking up, not noticing that he was a policeman. He called after me and gave me a jaywalking ticket for one hundred dollars. I apologized some more and tried to appeal to him by explaining that I had my PhD defense the next morning and was so focused on it that I hadn't noticed the crossing light. "Exactly," he said. "This will save your life. Next time you are so focused inside your head, you will remember that you are the only person in Milwaukee who ever got a hundred-dollar jaywalking ticket. Next time, you will pay attention to the traffic lights."

Things might have been different if I existed in a quantum world! In contrast to a classical object, which has a specific location at some point in space—for example, by the traffic light on one side of the road—a wave is an extended object that spreads throughout space. In this example, if I could have switched into a waveform, I could have existed on both sides of the road simultaneously—without breaking any traffic rules.

Subatomic particles are, to put it mildly, very different from large heavy objects (like humans), and they operate very differently; they are light, and they are tiny. It is in this domain that quantum theory rules and where all matter displays its dual wave-and-particle nature simultaneously. An electron, a proton, a neutron, a quark, an atom, a photon, and indeed the whole universe in its tiny infancy—all are waves and particles at the same time!

Ironically, we know this wave-particle duality is true from a nineteenth-century classical physics experiment first conducted in 1801 by Thomas Young. Called the double-slit experiment, it explains, very simply, a key property of waves known as superposition. When we have a bunch of waves in one place, they add up, literally, point by point; this addition is known as superposition of waves, and their pattern is called an interference pattern.

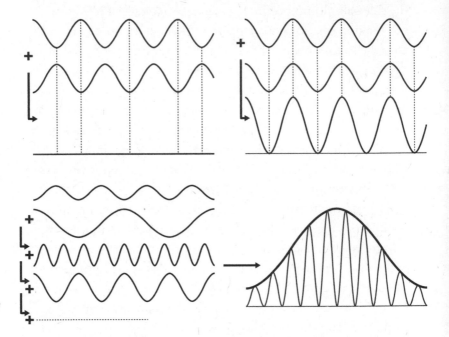

Figure 3. Waves' superposition when they are out of phase (*top left*) and in phase (*top right*). Many of them with different phases, frequencies, and amplitudes add up to a wave packet (*bottom*).

From daily experience, we know superposition to be true. When you attend a concert, you do not hear the sound wave produced by each individual instrument in the orchestra; rather, the music you hear is the packet of all the instruments' sound waves added together. Similarly, when you switch on the recessed lights in your hallway, you do not distinguish each unique light wave from each individual bulb; rather, you see the packet of light waves from all the bulbs blended together as one. The enveloping shape of all these superposed waves that move together as one unit is called a wave packet. (As we will see later, our universe at the earliest moments of its existence was a wave packet.)

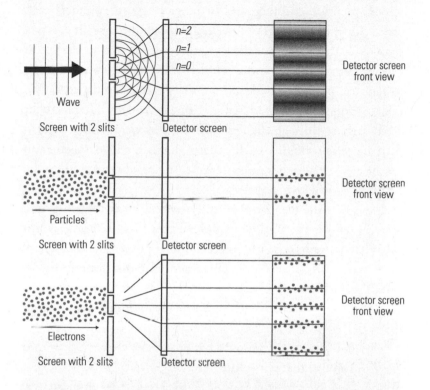

Figure 4. The double-slit experiment. *Top panel:* Light goes through two slits and creates an interference pattern on the screen behind. Middle panel: If electrons were particles, then we would see only two bright spots on the screen behind. *Bottom panel:* Electrons must also be waves because they add up and interfere, just like the light waves do in the top panel; therefore we see many bright and dark spots on the projection screen—the interference pattern of electron waves ("n" simply refers to bands).

In the double-slit experiment, the experimenter simply shines a light onto a board with two slits and then observes the pattern projected onto a screen behind. You can try this yourself if you are so inclined. The pattern that you will see on the screen is a series of alternating bright and dark spots, because in some places, the waves amplify each other, and in other places, they cancel each other out. If the light wave traveling through the first slit was at a peak while the light wave traveling through the second slit was at a valley, then the peak cancels out the valley; the two waves add up to zero and produce a dark spot on the screen, as in figure 4. (We call this "destructive interference.") But if the two different light waves passing through the two different slits were both at a crest, then the crests amplify each other—they are added together to produce an even higher peak—and appear as a bright spot on the screen ("constructive interference").

Generally, the bundle of sound waves at most locations in a concert hall or the bundle of light waves in your home is not completely in phase or completely out of phase but somewhere in between. Often the phases are randomly distributed; therefore, the addition of waves, rather than amplification of a flat line, leads to a wave packet with an enveloping shape, as in figure 3.

Moreover, it doesn't matter what kind of waves we use. Sound waves produce the same interference pattern as light waves. Sound engineers know that, due to destructive interference, there are areas of "cheap seats," where the music can barely be heard because when the sound waves reach that location, they are out of phase and cancel each other out. (These areas of the concert hall are, in effect, the dark spots from the double-slit experiment.) In the same hall, there are other areas of "expensive seats," places that benefit from constructive interference; here, the sound waves are in phase and the music is amplified. (These are the bright spots from the double-slit experiment.) Water waves produce the same interference pattern. If I were to throw two stones into a pool, I would see the resulting liquid waves meet and either amplify or cancel each other in modulations of troughs and crests, which form an interference pattern.

The same phenomenon happens in the quantum world, but with subatomic particles. When you shine a beam of electrons (or any other quantum particle) through the double slit, it gives the same interference pattern of dark and bright spots on the screen as the light waves. Indeed, the double-slit experiment offered a rare, early opportunity to test quantum mechanics. If quantum mechanics were nonsense, if there were no such thing as wave-particle duality, if particles were just particles, then sending a beam of electrons through a pair of slits would be akin to throwing marbles through open windows. The clunky marbles would leave random scratches on the wall where they hit. But the electron double-slit experiment reveals a complete interference pattern (the bottom panel of figure 4), confirming the wave properties of electrons. As we will see later in the book, when the waves are infant universes, the interference pattern of waves becomes pivotal in testing the origin of our universe.

Thanks to the work of Planck and de Broglie and the other giants of twentieth-century physics who followed in their footsteps, wave-particle duality is a fundamental concept in physics today—and it is also continuing to revolutionize our understanding of the cosmos. Perhaps nowhere is this revolution more profound than in the study of the universe's origins. We know that the infant universe—itself a quantum object—was a lot smaller than an electron or a quark. And as we will see shortly, quantum interference in the infant universe is a crucial key to unlocking the mystery of the universe's creation.

To the end of his days, Planck remained reluctant to take credit for the scientific revolution initiated by his own work. For more than two decades, Planck wished for his ideas to be considered purely mathematical rather than meaningful physical reality. His unease with quantum theory is best captured in his own words, engraved on the wall of the lobby in the Nobel Prize Museum in Gamla Stan, Stockholm. These words still ring true today: "A new scientific truth does not triumph by convincing its opponents and making them see the light, but rather because its opponents eventually die, and a new generation grows up that is familiar with it." (This saying has since been shortened to the pithier phrase "Science advances one funeral at a time.")

Nevertheless, Planck and other pioneering scientists had developed a new and lasting theory of nature. By the mid-1920s, the revolution started by Planck had become unstoppable. Previous phenomena that appeared mysterious in classical physics found simple explanations in the new theory. Bohr's atomic model and de Broglie's electron-as-a-wave model had pushed the boundaries further by showing that not only light but also matter is simultaneously waves and particles. Once convinced that they had the correct answer, quantum theory's founding fathers did the unthinkable: they remained steadfast against the might of classical physics and crossed over to the quantum realm, forever changing the way humans think. I would get a taste of what that kind of resistance felt like when going public with my own theory about the universe's origins—but I am getting ahead of myself.

Physicists, in my experience, have dual lives of their own. They can be regular people, relaxed, happy, goofy even, but they suddenly become the complete opposite when they are immersed in work or debating or scrutinizing each other's ideas. Time stops; life stops; there is no room for emotions. All that matters is mathematical rigor and razor-sharp logic, both of which require intense concentration. Solving the problem is the only thing that matters, because when you finally get the answer, it is magic.

My husband, Jeff Houghton, is not a physicist. I met him in Albania in 1992 when he arrived from Great Britain to work as an economic adviser in a European Union economic-development project, and we became friends. His project in Albania ended in December 1993, a month before I left the country to fly to the United States. Since we were going to be based on two different continents, I was convinced we would not see each other again. So he was the last person I expected to see on that January day a month later when I heard my name called on the loudspeaker in the Zurich airport and was instructed to approach the Swissair desk.

He was standing there in a casual manner as if meeting me by chance in the Zurich airport was a perfectly normal thing. He gave me a hug and asked if I would like a coffee, adding that he would like one because he had caught an early flight from London to Zurich in order to arrive before me. Then he asked if I'd like him to fly with me on the next leg of my trip to the United States.

I was confused, but to be polite I said, "Yes, please," which drew a round of applause from the people behind the Swissair desk. I asked him if we should get a plane ticket for him before we had coffee. The Swissair attendant smiled and said he had already bought a ticket for that flight, the seat next to mine, before my plane had landed.

For a while, before we married and had our daughter, he worked in Europe and I lived in the United States, but we spoke on the phone daily, thus managing to do what subatomic particles cannot: know with certainty both our speed and our location at the same time.

How scientists came to know about and describe the lack of certainty in the properties of the subatomic world—the next great "quantum" leap in our intellectual journey—is among my favorite stories in modern physics. It begins, appropriately enough, with someone who was essentially a graduate student.

In Copenhagen in the 1920s, a twenty-one-year-old highly gifted German student, Werner Heisenberg, attended one of Bohr's lectures on the atomic model and was so impressed that he asked to become Bohr's assistant. A few years later, in 1927, Heisenberg debuted the uncertainty principle, the central building block in the foundation of quantum mechanics—the theory on which I would rely in gaining a new understanding of the moment of creation.

Heisenberg's uncertainty principle declares that at the sub-atomic level, nature forbids us from knowing both the location and speed of a particle simultaneously with precision. It is central to explaining why the quantum world is filled with such constant uncertainty and why every outcome is based on probabilities. The principle is built on the dual nature of quantum particles as being both particles and waves. When we attempt to measure the speed at which the quantum particle moves, the particle switches to its dual twin, a wave. Instead of a point-like particle that inhabits a single zip code and address, the wave spreads throughout the universe, as illustrated in figure 5. Therefore, when the speed of the particle is known, its location is hard to pinpoint. And vice versa—when the location of the particle is measured exactly, then the range of possible values for its speed grows.

Thus, the speed and location of a particle are forever interlocked in a contradictory relationship. If you measure the speed of the particle with such accuracy that your margin of error is nearly zero, then nature forbids you from knowing precisely where the particle is located, no matter how clever you are and how sophisticated your measuring devices. The quantum particle (that is, the wave) may be found anywhere in the universe and you will never know for certain where. The reason behind this is explained by arithmetic: One divided by zero is infinity. If the error in measuring the particle's speed is nearly zero, then the

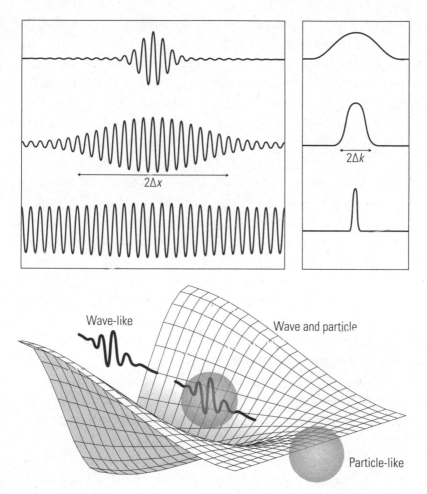

Figure 5. Top: Spreading of the wave and particle when the speed is known to great precision. A narrow wave packet (small Δx) corresponds to a large spread of wavelengths (large Δk); a wide wave packet (large Δx) corresponds to a small spread of wavelengths (small Δk). *Bottom:* An illustration of wave-particle duality set against an imagined space-time background. A particle is equivalent to the wave packet; most of it is "gathered" around the particle's location, but parts of the wave still stretch out to infinity.

error you will make in measuring its location is correspondingly very large—nearly infinite, in fact.

Heisenberg's uncertainty principle mathematically captures the uncertainty of the quantum universe, where information, be it energy or momenta, is spread over a range of possibilities instead of holding a single value. These possibilities do not exist in classical physics—in the visible world. But a quantum world is built from wave packets that wiggle around and spread out. The best we can do to describe a quantum world is estimate the chances (in science jargon, the probability) of each possible course that a quantum particle might take. And because our universe began as a quantum object, Heisenberg's uncertainty relations were intrinsically woven into its fabric from infancy and remain even in our big, visible, classical universe of today.

For cosmologists, the implications of Heisenberg's principle are simple: We cannot predict what will happen to the universe. The best we can do is calculate and speak of the probabilities that these events may come into existence.

Of course, Heisenberg was well aware of how weird his kind of universe sounded. As the British biologist J. B. S. Haldane put it: "The universe is not only queerer than we suppose but queerer than we can suppose."

Einstein himself could not digest the ludicrous implications of Heisenberg's principle. A quantum universe may sound weird even to seasoned physicists—but that is irrelevant. Whether or not we accept it, nature has chosen it for us; numerous experiments have confirmed the validity of quantum theory to a very high degree of precision. The universe has sided with Heisenberg.

But the consequences of an uncertainty principle when applied to the whole universe are alarming. They imply that nothing in a quantum universe can be determined with certainty. Ever!

Instead, they suggest that nature operates a lottery on a universal scale. If every possible universe in nature corresponds to an individual lottery ticket, each of them has nonzero odds of being the winning ticket, but none of them has a 100 percent certainty. Anything goes!

Imagine a universe that contains a planet Earth on which there is a little country called Albania. Let's say that 13.8 billion years after the Big Bang, there is a 30 percent chance that Albania will become a dictatorship, a 40 percent chance that it won't, and a 30 percent chance that this country may not exist in the universe at all. There is no way of knowing, at the moment our universe bangs into existence, which one of those events will be realized 13.8 billion years later. Instead, we have a pool of chances of various possibilities that an event may happen. Rather than operating in a deterministic way, every event in the universe, including the state of the primordial universe itself, is indeterministic. The universe is fundamentally uncertain.

Until the end of his days, Einstein was convinced that some profound insight was missing from quantum mechanics. He and other founding fathers of quantum mechanics could not accept the indeterminism introduced in nature by the uncertainty principle, so they tried to force their new theory into a construct that could support a single, deterministic universe. This is how the saga of a single universe continued throughout the twentieth century. But they failed. That failure eventually led to building the intellectual foundations on which the search for a testable theory of the multiverse began in the twenty-first century.

As if Heisenberg's uncertainty principle were not unsettling enough, a new development soon took the field of quantum physics in an even more uncertain direction.

Working independently of Bohr and Heisenberg in the early 1920s, the Austrian physicist Erwin Schrödinger, intrigued by de Broglie's findings that electrons were both waves and particles, focused on the wave-particle duality of quantum mechanics. Along with Planck, Einstein, Bohr, and Heisenberg, Schrödinger labored until the end of his life to disprove the implications of quantum theory.

Yet in 1926, Schrödinger made the most important discovery of his life and, unknown to him then, of the lives of generations of physicists. This discovery was the Schrödinger equation. It allows scientists to calculate what happens over time to a quantum particle as it is pulled or pushed by external forces. The Schrödinger equation also is the last pillar to complete our understanding of quantum physics.

Schrödinger's equation is not as impossible as it sounds. To make sense of it, imagine a group of physicists taking a walk on a mountain range—for example, the Rockies in the United States or the mountains of the Lake District in England. Suppose that the physicists are all carrying handfuls of marbles (the physicist's favorite toy!). At the top of a mountain, our physicists accidentally drop their marbles. In dismay, they watch their marbles roll down the mountain in all directions and settle into nearby valleys and lakes.

The marbles in this scenario won't stop rolling until they reach the bottom of each valley because the Earth's gravity pulls them down. In fact, they gain speed as they roll because their total energy (which is made up of their kinetic energy and gravitational potential energy) must be conserved. As the marbles get closer to the bottom, their interaction energy with Earth's gravitational field, known as potential energy, is converted to energy of motion, or kinetic energy, to compensate for the decrease in the gravitational potential energy, in order to keep the sum of the two unchanged. The mathematical expression for conserving the total energy is an equation that describes the classical motion of each marble as it rolls downhill.

Putting the difference between a classical and a quantum particle aside for the moment, Schrödinger's equation serves the same role as the above classical equation of motion for the marbles: Given the information on the mass of the particle and the external forces acting on it, this equation describes how a quantum particle evolves in time. It is the quantum version of the classical equation of motion. Both sets of equations are guided by the same principle: they constrain the total energy of the particle—the sum of its kinetic and potential energies must be conserved. Energy cannot be created out of nothing; thus, it must be conserved.

To see how this works, imagine now that Earth's gravitational potential energy represented by mountain ranges and the marbles have been rescaled to be subatomic size—say, the size of electrons. These electrons are in the presence of some external field similar to the gravitational field on the marbles in a mountain range. The type of the external field to which these electrons are subjected does not matter for our purposes; this external force could itself be gravitational, nuclear, electromagnetic, or something else. All that matters is that there is an external force exerted on the electrons, just like the gravitational pull of the Earth is exerted on the marbles. The motion of these electrons is given by solutions of the Schrödinger's equation.

Despite their similarity, the classical and Schrödinger equations of motions are conceptually very different on some key points. Schrödinger's equation operates in a quantum world, and it treats particles as if they were waves. To complicate matters, it doesn't produce a single answer for how a "quantum marble" will move. Instead, it offers a family of waves with each of the waves moving on a different path.

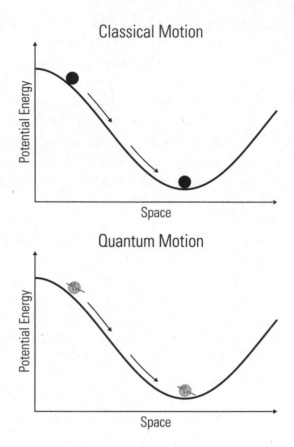

Figure 6. The top graph shows a marble rolling down a mountain under the influence of the Earth's gravitational potential energy. The bottom shows a quantum particle, such as the inflaton, rolling down a potential energy field. Thanks to Schrödinger, we know that this quantum particle is also a probability wave.

Crucially, each wave solution coming out of Schrödinger's equation is interpreted as a probability wave, meaning a quantum particle has a nonzero probability of following any of these wave paths—and we won't know beforehand which one the particle will choose.

Furthermore, unlike visible, classical particles, which follow determined trajectories in real space-time, quantum particles evolve in a space of possible paths, each with its probability of happening.* That multitude of possible paths captures the full uncertainty of the quantum world.

Heisenberg, the originator of the uncertainty principle, approved of Schrödinger's wave solutions being conceived of as probability waves.†

With Schrödinger's equation, the complete quantum "machine"— the mathematical device—that nature uses to run its probability game was finally revealed. Into one side of the equation, we feed the mass of the quantum particle and the external forces acting on it; on the other side, the machine spits out the answer, the probabilities for what may happen to the paths that particle will follow over time.

---

* The collection of these probable paths of waves is called the Hilbert space, named after the mathematician David Hilbert.

† To Heisenberg, the abstract space where quantum particles operate, the Hilbert space, and Schrödinger's equation, which shaped the evolution of a quantum world, were representations of Plato's abstract double levels of existence, the physical world of space and time in which particles and objects exist, and a higher level, after which the world was shaped, the "demiurge": "Modern physics has definitely decided in favor of Plato. In fact, the smallest units of matter are not physical objects in the ordinary sense; they are forms, ideas which can be expressed unambiguously only in mathematical language." See Werner Heisenberg, *Das Naturgesetz und die Struktur der Materie* (Germany: Stuttgart Belser, 1967), 68.

So what happens if we turn this quantum machine to the seemingly intractable problem of the explosion of the early universe? What happens if the particle that we feed into the machine is the inflaton, the cosmic particle that drove the inflation of the universe at its infancy? What insights can quantum physics provide about this mysterious moment? How can it help us understand the elegant but incomplete theory of cosmic inflation?

When Guth and Linde's inflaton particle and its hypothetical potential energy are fed into the machine of the Schrödinger equation, it spits out the answer of an inflationary universe. Much like the marbles rolling down Earth's gravitational potential energy on the mountain, Guth and Linde's inflaton rolled down a potential energy well that was so shallow, the inflaton rolled extremely slowly—so slowly that its energy didn't appear to change over time.

Cosmic inflation is a paradigm rather than a theory because there are many models, postulated by many scientists, for what the hypothetical primordial potential energy of the inflaton could be. Despite the variety in their postulated inflation energies, all of the models of the cosmic-inflation paradigm achieve the same result: the inflaton breathes its fire into the Big Bang, "blowing up" the infant universe and accelerating the universe's growth uniformly in all directions. A little later, this same quantum particle fills the universe with all matter and light and energy to make it into the beautiful place we call our universe today.

But how do we go from the origin of a quantum particle to a vast collection of matter, stars, galaxies, and planets—a universe so expansive that we could not journey across it even if we were given many lifetimes? Here, the theory of cosmic inflation seemed to come up short—which intrigued me and made me wonder what path Boltzmann, Planck, and Penrose would have chosen.

4

Fine-Tuning

P HYSICISTS ARE HAPPIEST when they identify new problems rather than new solutions. (Sometimes also when they are told that they are smarter than their colleagues.) New problems mean the potential for new discoveries and a subsequent lack of boredom. And such is the beauty of the quantum world that it never ceases to offer new insights into old problems, including the rich story of cosmic inflation.

After receiving my doctorate from the University of Wisconsin–Milwaukee in 2000, I decided to devote most of my time to cosmology. Instinctively, I knew this field was about to enter a golden era. Technology was allowing us to explore farther and deeper into outer space and microscopic space alike, and the standard model of cosmology—the dominant theoretical model that describes the evolution of the universe—contained more mysteries than ever, mysteries that a new generation of scientists would have an opportunity to solve.

And as I embarked upon my newly chosen scientific career, one mystery in particular captured my interest.

• • •

In the standard model of cosmology, a minuscule fraction of a second after the Big Bang ends, the universe is roughly the size of a blueberry; it has exited the initial inflationary stage. It is a classical object (although the stuff inside it is still quantum), and while its growth is no longer accelerating, it continues to expand. But if nothing can exceed the speed of light, how do the wavelengths of these particles catch up and communicate with each other over the growing distances across the expanding universe?

Here, the wave-particle duality of quantum mechanics might provide the explanation. As the primordial patch of space is blown up like a balloon in the same manner in all directions to cover the whole infant universe, everything inside is being stretched along with it.

To be clear, while particles are assumed to be stretching at the same scale as the universe, this does not mean that an electron suddenly becomes as big as the whole universe. Rather, what changes is the particles' wavelengths. These particles are still sub-atomic, governed by quantum theory, and thus retain their wave-particle duality. So, as the universe inflates, these photons' and particles' wavelengths get stretched, not the particles themselves.

This phenomenon accounts for something else: the cooling of the infant universe. As the wavelengths of these wave-particles expand, their energies decrease in an identical relationship, which causes the whole universe to cool down uniformly as it expands.

The temperature of the universe continues to drop after inflation ends, since the universe continues to expand. But when inflation ends, the energy of the inflaton has to go somewhere— and it does. The inflaton particle decays into other particles of matter and light and transfers all of its energy to them, a process known as reheating. (In truth, the universe does not actually reheat; rather, it is filled with radiation, just like George Gamow predicted in his Hot Big Bang model. The word *reheat* is a misnomer derived from the explanation for the similar properties of our universe after inflation and Gamow's Hot Big Bang.)

This continuous cooling as the universe expands creates the conditions for elementary particles—protons, neutrons, and electrons—to become stable. They first appear at about a ten-millionth of a second after the Big Bang. A bit later, these particles combine into the first atoms of hydrogen (see figure 7), and the universe is covered by a cloud of hydrogen.

Three minutes into the life of the universe, a phenomenon known as Big Bang nucleosynthesis (BBN) takes place. BBN produces helium and other elements heavier than hydrogen that populate the universe, and it lasts for about four minutes. At this stage, although the universe has begun to cool (it technically started cooling almost as soon as it began to inflate), it still remains so hot that photons—that is, light—and matter particles like protons, neutrons, and electrons are boiled and mashed together into a plasma state, making the universe opaque.

Not until some 380,000 years later does the universe's temperature fall sufficiently for photons to completely free themselves from other particles in the plasma; they thereafter remain visible to observation as a bath of radiation in the background sky. As these light particles separate from matter, the universe becomes transparent. This background bath of radiation, which continues to permeate the outer reaches of the universe to this day, is called the cosmic microwave background radiation, or CMB. (I will return to CMB later on, because it is going to matter a lot as my own quest progresses.)

A few billion years or so after cosmic inflation ends, structures (stars, galaxies, and clusters of galaxies) start to form, the result of the gravitational condensation of the remnants of the hydrogen cloud. All the regions in the sky that have high concentrations of matter collapse gravitationally under their own weight and start making the first stars. The next stage is the production of heavier elements. Metals are produced by fusion—the smashing together of lighter elements at extreme pressure to make a heavier one—inside the cores of stars.

Once stars form, the universe goes through a long and perhaps less interesting period where nothing major happens; the universe continues to expand and its temperature continues to cool. (The reconstructed history of our inflating universe in space and in time from the first moment to the present is shown in figure 7.)

But if our universe started purely from the energy of cosmic inflation, how did everything within it, from photons to matter particles that make stars and galaxies, come to be here?

Cosmic inflation has an answer to this question, too, and a very good one. While Einstein's equations convincingly demonstrate that the overall growth of the universe comes from the matter and energy inside it, quantum theory reveals what seeded this matter and energy in the universe when cosmic inflation ended. The origin of these structures arises from Heisenberg's uncertainty principle. In quantum mechanics, fluctuations of energy, including the energy of cosmic inflation, are always present. We can think of quantum fluctuations as unpredictable small deviations that flicker on the paths of quantum particles and as variations that glint their energies. They are mathematically captured by the Heisenberg uncertainty principle, therefore, since quantum fluctuations in the initial energy of inflation are unavoidable, then when the universe stops inflating, it suddenly finds itself filled with waves of quantum fluctuations of the inflaton energy. The whole spectrum of the primordial fluctuations in the inflaton energy, some with mass and others without, are known as density perturbations. The shorter waves in this spectrum, those that fit inside the universe, become photons or particles depending on their mass as the universe cools down.*

After inflation ends, the primordial soup of photons and particles is scattered uniformly through the universe as a distribution of blobs (lots of mass) and voids (little mass). The over-dense regions with blobs of mass collapse under their own weight and create stars and galaxies. The sky tonight is the result of these primordial fluctuations; they show up as light and stars. The origin of matter and light inside the universe is, therefore, purely quantum in nature.

Furthermore, Einstein's equations connect the energy of these fluctuations to the space-time on which they exist (remember, energy tells space how to curve). Specifically, the energy contained in the inflaton fluctuations triggers tiny tremors in the very fabric of space-time. And the tiny tremors in the fabric of the universe in turn induce weak ripples or vibrations in the gravitational field inside the universe. These ripples are known as primordial gravitational waves.

We can make models of failed universes. Different inflationary models produce different amounts of matter and light, resulting in different density perturbations. These density perturbations, which establish the matter-energy content of the universe, determine what happens to its growth. If a lot of matter is produced, the universe will not hang around long enough for humans to arrive; it will quickly cave in on itself and collapse like a black hole when it is still young. If there are too few of these fluctuations,

---

* Based on the principle of wave-particle duality, we can think of the fluctuation waves as particles. In the dual picture, the massless particles become photons, and the massive ones become the normal matter particles that make stars and galaxies and us.

the universe won't have enough matter to clump. The over-dense regions will be too few and too far apart—with the result that the universe will be barren of life and structure. It will continue expanding but will be comparatively empty.

In my mind, the answer to this riddle—the extremely small chances that our universe has to start with the right type of inflaton potential energy and a smooth patch of space—lay in decoding where the inflationary energy originated and what was there before.

The strength of the inflaton fluctuations that create the desired amount of density perturbations and, ultimately, all the matter and radiation we observe in the universe are directly determined by the details of the energy of the inflationary model. The trouble is that there is a large family of models under the umbrella of the cosmic inflation paradigm. Choosing the right potential energy, the one that contains the desired features that reproduce the structures we observe in the universe, is a theoretical construction done by hand, a model selected from a myriad of potential inflaton energies. Physicists hoped that one day, instead of an ad hoc design, they could motivate and derive this unique model of cosmic inflation based on fundamental physics laws. This challenge intrigued me.

As I went through a group of popular models of cosmic inflation, I could appreciate what made the opponents of the theory most unhappy: If the inflationary model had to be meticulously constructed to give the right answer—that is, the right amount of perturbations to explain how our universe came to exist in the exquisitely balanced form that it did—then isn't cosmic inflation an unnatural way of starting a universe? Observationally, our universe does turn out to have the perfect amount of density perturbations (about one part in a hundred thousand) to stay flat and hang around long enough for stars to form and make heavier elements and—eventually—human life. The opponents of cosmic inflation were rightly concerned that the hypothetical potential energy and the initial tiny patch from which our universe inflated appeared carefully designed, or fine-tuned, to produce precisely this amount of perturbations.

This special arrangement seemed to be the crux of the origin problem. If the inflationary universe is in a specially ordered state at its initial moment of existence, then its entropy must be nearly zero, which means it has an infinitesimally small chance of happening. In other words, for cosmic inflation to switch on and kick-start the universe into being requires very special initial conditions indeed. To have all of these—a flat universe with the right amount of structure scattered uniformly and homogeneously through our skies—for the price of one inflaton particle, the infant universe must have been in a remarkably unusual state of exceptional order.

Here is the dilemma physicists face: Cosmic inflation offers the whole cosmic origin story in one irresistible package. But it does so at the cost of one assumption: a finely tuned start of the universe at high energies on an exquisitely smooth tiny patch of space. And this is a huge assumption, because everything else we know about the workings of the universe tells us that the odds of our universe starting the way it did, in a tailor-made initial state of exceptional order with an entropy state of nearly zero, are ominously small!

Penrose had made a splash in the 1970s by pointing out this embarrassing fact, and it led to some even more embarrassing implications. Since starting a universe with this state is more improbable than any other possibility, then even what might be conceived of as outrageously impossible will have a higher chance of existing than our universe does.

Consider this spooky example as a dramatic statement of the un-likeliness of our existence: The spontaneous formation of a brain in empty space stands a much greater chance (statistically) of occurring than the creation of our universe through cosmic inflation! I kid you not. This compelling description of improbability has become known as the Boltzmann brain paradox in celebration of the legacy of the man whose own mind produced the entropy equation and its relation to probability. The standard model of cosmology does indeed seem to lead to the conclusion that floating brains and all other sorts of science-fiction events you never thought possible should not only exist but outnumber and overwhelm us. As absurd as this floating brains idea sounds, you won't be able to get a straight answer from any physicist as to why these brains are not there. Granted, it is an outlandish and ludicrous example, but it provides a dramatic indicator of the extreme unlikelihood of cosmic inflation.

For some scientists, these were good enough reasons to discard cosmic inflation and replace it with a new model of cosmology. While open to persuasion, however, I still believed that cosmic inflation was correct but incomplete.

Moreover, as I spent years going through every page of our cosmic evolution, I became convinced that discarding cosmic inflation was not a wise solution to its origin problem. Cosmic inflation has performed spectacularly well against all observations that have been made of the universe to date.

Meanwhile, the pressure from freshly discovered cracks and paradoxes in our understanding of the universe was growing. In addition to the mystery of its origins, the universe was about to throw another curveball at us. In 1998, while I was still a graduate student at UWM, the distinguished astrophysicists Saul Perlmutter, Adam Riess, and Brian Schmidt had made a surprise announcement: The universe was inflating—again!

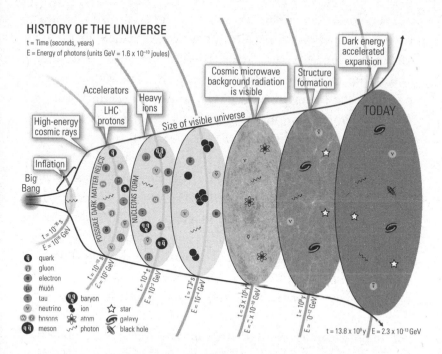

Figure 7. The whole standard model of cosmology is contained in this diagram. Time runs on the horizontal axis from the first moment of Big Bang inflation to present. Space runs on the vertical axis. Note that the universe is growing in time and it is flat in space, as can be seen by taking slices of the universe at each moment in time.
Particle Data Group at Lawrence Berkeley National Lab

The name for the energy that is making the universe inflate again is mysterious and foreboding: dark energy. Dark energy is a true enigma even to the best scientific minds. It became my particular fascination by the end of my graduate work; indeed, you could say that my deep interest in the universe's beginnings was cemented by contemplating how it will end. During my two-year postdoctoral fellowship in Pisa, Italy, at the Scuola Normale Superiore di Pisa, that was my sole concentration.

Scuola Normale Superiore is one of the most inspiring institu-
tions I have ever worked at. Its rich history provided the perfect
setting for contemplating the destiny of the universe. Galileo
and, more recently, the great nuclear physicist Enrico Fermi
had trod the hallowed halls of the university. Fermi's former
office was only two doors from mine; I could see the Leaning
Tower of Pisa from my window.

Dark energy behaves like an ether; its energy mysteriously
diffuses out of a vacuum and yet it permeates every speck of the
universe and pervades the very fabric of space-time. Similar to the
energy that jump-started cosmic inflation, dark energy has two
unusual properties: the amount of it per unit volume, its energy
density, is (almost) a constant, and its pressure is negative. Both of
its components, energy density and pressure, will determine what
happens to our universe in the future. While the total amount of
dark energy contained in the universe controls the speed of expansion of the universe, its pressure controls the acceleration of its
expansion—that is, the rate at which the expansion of the universe
increases or decreases.

The total amount of dark energy grows in direct relation to
the expansion of the universe, so as the universe expands and
every other energy source dilutes and ultimately empties out,
dark energy will be the only energy left. Thus, dark energy becomes the ultimate arbiter of the universe's destiny. The density
of dark energy is tiny—approximately one one-thousandth of
an electron volt. But while this number may sound small, don't
be fooled: dark energy dominates the energy of our present
universe.

Visible matter, or what physicists call baryonic matter or
baryons, is the stuff we are made of; it includes every proton and
neutron, all the atoms in our bodies and in every living thing,
the stars and planets, the galaxies and clusters of galaxies, and the

cosmic dust. In sum, it is all the material that we *see* in the world around us—and it accounts for less than 5 percent of the total energy density in the universe. Dark matter—a type of matter that is nonluminous and therefore invisible to the eye—makes up about 20 percent of the universe's energy. The remainder, dark energy, makes up a staggering 75 percent.

Oddly, our universe had the perfect amount of dark energy at the beginning—only a tiny bit—which allowed the cosmos to hang around long enough for all the structure and life to occur. The baffling question, one for which no one has an answer, is: Why isn't this number zero or, alternatively, the same as the Big Bang energy?

The discovery of dark energy presented two other daunting mysteries. First, if our universe had contained even a little bit more of this unfathomable energy, it would have grown too quickly and too early, before matter clumped together and formed stars. The universe would have been ripped apart and would have remained featureless from the accelerated expansion billions of years ago. How did we get so lucky?

Second, just like the mysterious source of the energy that unleashed cosmic inflation at the beginning of time, the origin of dark energy is unknown. While I was in Pisa, my Spanish-born colleague and friend Mar Bastero-Gil and I proposed a likely source for dark energy, arguing that it was stored in a particular way in the quantum fluctuations of what might be empty space-time or, perhaps, a vacuum. But as of this writing, this second mystery endures too.

Without the answers to these questions about dark energy, we cannot definitively settle the larger question of how the universe will behave in the future. It is littered with Heisenberg uncertainties—which might be just as well, because the options are not encouraging.

What might a dark-energy future hold? According to what we do know about it, dark energy promises multiple cataclysmic endings for the universe, more shocking than even the most inventive science-fiction writers could imagine. Soon (and don't worry—by *soon*, physicists mean cosmological timescales measured in billions of years), when dark energy is virtually the only energy left in the universe, galaxies might break apart so quickly that they will lose contact with one another. Each local galaxy, including our own Milky Way, would become its own universe, disconnected from the other regions and separated by distances that can be reached only at superluminal speeds.

The universe's ultimate death could, cosmologists hypothesize, come in one of at least three ways. If our universe remains in an accelerated expansion forever, the distance between stars and galaxies will be so big that our sky will appear empty, and the universe's temperature will fall to nearly absolute zero. Life will not be able to arise or be sustained; starlight will not reach us; the sky will be frigid, empty, and dark. All species would suffocate; all clocks would freeze. The last tick of our cosmic clock, just before the end of time, would become the universe's last heartbeat.

A second possibility is that dark energy is not actually vacuum energy but an inflaton-type particle that temporarily moves so slowly that it mimics pure vacuum energy. But in the future, this particle may decide to alter its speed and energy. If it speeds up, it could ultimately reverse the universe's expansion and cause it to contract and get hotter and then collapse in a blaze of fire.

Finally, dark energy could become what has been described as phantom energy, energy that becomes wild and uncontrollable. In this scenario, galaxies and stars and atoms and, ultimately, the very fabric of the universe are ripped apart. Worse still, the process is accelerated, and the universe will be shredded in a relatively short time, perhaps only ten to the power of thirty billion years.

Of course, the fate of the universe remains to be written. But what we do know is that our ending is inseparably intertwined with our beginnings, because it was inflation energies and dark energies together that drove our universe into its accelerated expansion. Both have and will control the growth of the universe from its first moment to its last. And from the start, both had to be calibrated just right to ensure that our universe survived. Otherwise, matter could not have clumped, stars would not have formed, we wouldn't be here, and the universe's very fabric would have already ripped apart billions of years ago.

The revelation that both the origin and the destiny of the universe rely on the same underlying enigma, that the same type of energy that sparked the first episode of inflation and created the universe will also produce the final episode of inflation when the universe ends, led to scientists' renewed interest in our origins. Adding dark energy to the mix of problems associated with cosmic inflation also raised a challenge to the universe's origin story that was too important for physicists to ignore.

In the early 2000s, as scientists were faced with these two fundamental mysteries—the creation and the final destiny of the universe—cosmology entered a golden era: an age of big problems in need of solutions. And as luck would have it, it was also when I began my career as a full-fledged working scientist. Like my colleagues, I couldn't wait to get busy.

Physicists are very good at playing God using so-called thought experiments. We cannot reproduce the Big Bang inflationary explosion to create universes in a lab, nor can we travel back in time to explore what was there before the Big Bang. However, our minds are our cosmic labs; we can use the tight constraints of math, the laws of nature, and our astrophysical observations to imagine, scrutinize, and sift through all the possible scenarios in the form of thought experiments. So, as is customary for a "pen-and-paper" theoretical physicist, I started by coming up with a variety of new thought experiments and analyzing existing ones.

In my first thought experiment regarding the universe's creation, I relied on the Penrose theorem that the entropy of the universe is quite large at present but was nearly zero at its first moment. I envisioned starting with a big universe full of entropy and with all the structures already in it, like our present universe. Wouldn't that be far more likely to happen than starting with a tiny universe that banged into an explosion and went through all the trouble of actually making the structure of stars and galaxies (and us)?

I tweaked this example further. What if I were to conjure up another non-special universe, one that started big but, instead of expanding, kept shrinking until it crunched to a point? In this new universe, I would have naively reversed the direction of time (our second law of thermodynamics), exchanged the future with the past, and swapped the present high-entropy state of the universe with its past low-entropy state 13.8 billion years ago.

Could these tricks solve the origin problem? Is a universe starting in a state of high entropy more likely to occur than the real universe we inhabit? Unfortunately, no—it is not that easy! This new universe that started big and full of entropy turned out to be just as unlikely to occur as our universe, which started small and devoid of entropy. For one simple reason: according to the second law of thermodynamics, the entropy of a universe keeps increasing relative to where it began. As we discussed in chapter 1, Boltzmann taught us that the entropy of the universe is a measure of its probability to spontaneously come into existence; therefore, if the universe had a low entropy at its creation relative to the present, then, correspondingly, it would have a low chance to exist.

Fine, I thought, *the new universe model doesn't solve the problem of the unlikeliness of our origin.* But how about another possibility: A universe that bounces back and forth, from big to small and from small to big again, in eternal cycles? That should evade the second law of thermodynamics, since we have completely lost track of what to call the beginning of the universe and what to call the end. When the universe bounces and repeats its bouncing, all the cycles are the same. Such a universe should give us the freedom to choose an initial state of high entropy and a final state of low entropy. We can choose the beginning of the universe to be the moment when it has just crunched at the end of one cycle, the moment when it has just bounced back and reappeared small and growing in the next cycle, or the moment when it is halfway through the cycle.

Such models do exist. But like the previous one, they collapse in humiliation thanks to the second law of thermodynamics. The entropy of any universe we model cannot decrease with time, ever. That is, a universe simply cannot reorganize itself to go spontaneously from a disordered state to an ordered state. What I learned from playing out such thought experiments in my head was that it didn't matter how we started or what kind of a universe we have; we will still conclude the origin of that universe is special. Our universe was not unique in starting with a low-entropy state and therefore a low probability of existence. The second law of thermodynamics made *any* universe appear special.

Although this way of thinking took me further away from a solution, I realized that existing models built on such reasoning were too naive. There is no such thing as an absolute measure of entropy in the universe; an entropy state is high or low only relative to some other entropy state. All these models failed because

I applied the second law of thermodynamics to a single universe. Whichever new universe I started with, its entropy would increase in the future relative to its own entropy at the beginning.

No matter what we call the first moment in the life of the universe, the universe will increase its entropy the instant after it begins and will continue to do so in the future. As entropy increases, disorder and the amount of "missing" information increases over time (recall that, according to Boltzmann, entropy is simply the missing information about a microstate of the system contained in any given collection of microstates). That means that no two cycles in the bouncing universe can be identical; each cycle will have its own individuality. Building a universe that goes through identical cycles is impossible, since it keeps increasing its entropy over each cycle. The universe cannot recover in one cycle all the information it lost in the previous cycle; the lost information, in the form of entropy increase, is lost for good. Entropy increases irreversibly from past to future. Thus, the cycles of this universe cannot be identical or reversible; they cannot evade the second law of thermodynamics or the irreversible arrow of time (the direction of time from past to future).

With the help of these thought experiments, I had reached a turning point in my investigation.

I concluded that, quite generally, the origin of any single universe is unlikely, no matter what people conjured up in thought experiments. Independently of how the universe starts and how it evolves, whether it grows, shrinks, bounces, or gets ripped apart in its final state, it will always evolve toward disorder; its future will always be more disordered than its past, because its entropy must increase. Which is why the second law of thermodynamics that states "entropy always increases" holds a supreme position among all the laws of nature. These thought experiments helped me realize that the mystery of our unlikely existence is a generic problem for starting any universe in any manner. Which didn't make sense.

This reasoning convinced me that deconstructing and reconstructing the universe would not help shed light on its origin. No matter what kinds of universes I made up in my mind, they would all suffer from an unlikely origin. The constructs were doomed. But at least they helped me narrow down the possibilities and figure out what didn't work.

The common thread in these thought experiments was that they were based on comparing entropies within a single universe. This made me think that the hypothesis of having only one universe was why these thought experiments ran into trouble with the second law of thermodynamics and therefore always failed to explain the origin of the universe. The single-universe assumption was to blame. Why were physicists still clinging to it, I wondered—and what would happen if we discarded it?

It is difficult to overestimate the appeal of a singular universe to most twentieth-century physicists. The beauty of science lies in its simplicity, a simplicity encoded in the logical structure of its equations. At the same time, the value of science lies in its predictive powers, in its ability to state with certainty what will happen to an object—in this case, the whole universe. The idea of a single universe described by a single unified theory provided both simplicity and predictability. It satisfied both of these basic scientific cravings.

The bias toward a singular universe with a single set of laws ruling it is an ancient idea dating back to Plato. Much more recently, Einstein spent the later years of his life in search of a single theory of everything that would reveal the one manual of laws covering our whole universe, from its origin to its ultimate destiny.

As I slowly chipped away at the problem of the universe's origins, I came to see that most of the prominent ideas circulating in the physics community were not fundamentally different from earlier ideas going all the way back to antiquity. And previous attempts had failed to solve the problem. Before digging too deep, I thought it was important that I understood other researchers' points of view. Specifically, what made these scientists keep returning to the same framework of a single universe wrapped up in a theory of everything or, alternatively, steering away from this problem altogether? I suspected one of the reasons was the long history of what I call the single-universe school of thought.

Although I still had no clear idea of what might work, I realized that this mystery required a different approach. If the small probability of existence was generic to all universes, then something very basic must be missing from our understanding of the issue. What could that be?

5

Are We Alone?

THE CHARGED ATMOSPHERE in the theoretical physics community at the start of the new millennium eerily reminded me of the desperation and uncertainty that my family experienced when my dad was exiled for the second time. Back then, another aftershock from the cultural revolution that had been imported from China in the 1970s was shaking Albania. My dad was working as a scientist at the Albanian Academy of Science, and his research results on the optimization of forestry, hydropower, and other difficult problems based on statistical probability were sufficient to draw the jealousy of some powerful colleagues. Once again, my dad's enemies could exact their revenge by targeting his weakest point: his "bad biography," our family history.

This time, my father was sent to work on a cooperative in Dukat, a remote mountain village in the south of Albania. On that particular occasion, my dad's concern for our safety led him to embrace a lie, even though it would cause him great personal hurt. Such was the Albanian reality then.

During my dad's second exile, my mom's file—every Albanian had a file cataloging his or her life and behavior—was taken from the League of Writers and Artists and sent to the isolated mountain village where my dad had been banished. For months, my mom had been pressured every day at her office in Tirana to pack up her belongings, take her family, and follow her husband to the remote countryside. After her file was transferred, the request for her to join her husband was elevated to an official order. Terrified at the idea that his wife and two children would also be sent into exile, my dad begged my mom to denounce and divorce him, for show. It was the only way for her to distance herself quickly from the "enemy of the party" (my dad) and keep my brother and me safely in Tirana. But Mom wouldn't have it, so my dad sent my maternal grandmother to Tirana to talk sense into her daughter. Eventually, the two of them wore down her defenses, and she agreed.

On one of my father's brief visits home, my parents, grandmother, and I went to the courthouse, having agreed beforehand that my mother would lie and claim that my father had beaten her. It was the fastest way to get a divorce. A female judge was assigned to the case. My mother entered her chambers and said, "My husband beats me. He is here; you can ask him yourself. I want a divorce."

But what had seemed like such a simple solution broke down the moment Mom uttered those words. It turned out that the judge was a childhood friend of my father's, and she adored him. She was furious. Staring down at my mother from the bench, she thundered that she knew my father's character, and he would never hurt anyone. "What kind of a woman are you? That man wouldn't even hurt a fly!" the judge yelled. "These are vicious lies. How dare you? Do you understand what kind of trouble you can get him into with your lies, as if he isn't in deep enough trouble already? Get out of my sight!"

My mother, who had not wanted to agree to this plan in the first place, fled the court in tears, and we had to run to catch up with her. The divorce scheme failed; the truth of my parents' own small universe won out. My dad was returned to Tirana a few months later, and our family stayed together.

Such memories crossed my mind years later as I started to suspect that a drastically different way of thinking was required to solve the question of how the universe had been created. What if the thing we saw as the problem—the universe's improbable creation story—was not, in fact, the true problem? I suspected that part of the solution was hidden in the paradox that I had found in my many thought experiments: no matter how many different types of conceivable universes I could possibly imagine, the problem of an exquisitely special and extremely unlikely origin was the same for each of them. Did all of these universe models, no matter how different, share the same logical flaw?

Very soon, I had an idea. But to pursue it, I would have to break away from the scientific community's orthodox picture of a single universe described by a unified theory. And in the world of twenty-first-century physics, that amounted to nothing short of heresy.

I started researching the origin of the universe full-time when I arrived at the University of North Carolina at Chapel Hill in 2004. I had no preconceived idea of what the ultimate theory of the universe should look like, which worked in my favor. I also had no preconceived idea of where my investigation of the origin of the universe would lead.

What I had realized by this time was that every model and every calculation, whether promising or a failure, started with the same fundamental assumption: *that there is only one universe.*

Perhaps, I thought, by refusing to allow for the existence of anything other than our one universe, physicists had already made it "special." Perhaps, I thought, we had been asking the wrong question all along. After all, how could we ask ourselves why we started with *this* universe if all we allowed ourselves to consider was a *single* universe?

I gradually convinced myself that the question "Why did we start with this one?" logically made sense only if we had a range of possible beginnings to choose from—a range of different infant universes, any one of which might have turned into the universe that we find ourselves inhabiting today. Otherwise, the answer would be obvious: We got this universe because it's the only one there is. End of story.

From the beginning, I expected that the idea of starting with multiple infant universes was going to be an unpopular, if not an outlandish, premise. Since antiquity, the philosophical concept of a single universe has dominated and shaped nearly all efforts to understand the universe. In modern times, this belief had crystallized into a global scientific effort to unify quantum theory (which governs the microscopic, unseen universe) and Einstein's theory of curved space-time gravity (which governs the visible universe) to produce a theory of everything—a scientific model capable of explaining all the workings of our universe. Indeed, it is possible to draw a nearly uninterrupted line from Plato and Aristotle directly to Einstein and Stephen Hawking in support of a theory of everything for a single universe. To go against this vision was to go against the titans of philosophical and scientific thought for three millennia.

I wasn't the first one to encounter this problem. The founding fathers of quantum mechanics got the same answer I did, a family of universes instead of a single one, when they solved the Schrödinger equation. Nevertheless, some tried their best to force their new quantum theory to produce the answer of a single, deterministic universe. And they did that in some very creative ways.

Recall, for example, that the solutions of Schrödinger's equation showed that a quantum particle can follow multiple paths, and we won't know beforehand which path the particle will choose; each has its own probability of happening. If the quantum particle happened to be an infant universe, then a family of solutions implied the possibility of a family of infant universes, each with its own probability of existing, because, on the basis of a wave-particle duality, each wave solution could be thought of as its own infant universe. Since each wave solution corresponds to an infant universe with its own probability, from now on in this book I will refer to them as wave-universe solutions. The collection of all these wave-universes derived by solving the Schrödinger-type equation for the universe is known as the wave function of the universe, and each individual wave-universe is a branch in the wave function of the universe.

For the founders of quantum theory, then, the question became this: Which one of these wave-universe solutions was the real one? Simply ignoring and discarding all the solutions save for the one that they liked and presenting that as the correct one seemed outrageously arbitrary. However, nobody knew what it meant to keep them all or how to identify which one of them would become the "real" universe because, as it soon became clear, Schrödinger's equation and Heisenberg's uncertainty principle didn't offer a selection criterion that singled out one "valid" infant universe among of the multitude of their solutions.

Niels Bohr was adamant that somehow a *single* large universe had to be identified out of the family of quantum baby universes produced by the Schrödinger equation. If not, all the predictability of physics seemed poised to be lost. But how to identify the winner of this cosmic lottery? Bohr proposed bringing in a hypothetical independent judge, someone who could observe the wave function of the universe and rule in favor of *only one* probability wave out of the multitude. Bohr argued that the moment the judge observes what course a quantum particle has taken, then we can know with 100 percent certainty that this particular particle is the real one in the family of wave solutions because we have just observed that it exists. And we can then discard the rest of the possible answers.

Bohr's solution became known as the collapse of the wave function because all but one branch from the family that makes up the wave function of the universe vanished. In this manner, the whole wave function was reduced; it "collapsed" from a multitude to a single choice, the surviving branch. To many, it sounded plausible. Indeed, Bohr's collapse of the wave function dominated theoretical physics for decades. It still has a few fans left today in the physics community. I, however, am not one of them, for reasons that I will explain.

Heisenberg, for his part, was on board with most of Bohr's proposal. Indeed, the creator of the uncertainty principle is credited with naming Bohr's contribution to quantum physics; Heisenberg called it "the Copenhagen interpretation of quantum mechanics." As Heisenberg said in a 1924 lecture in Chicago, "The probability wave meant a tendency for something. It was a quantitative version of the old concept of potentia in Aristotelian philosophy. It introduced something standing in the middle between the idea of an event and the actual event, a strange kind of physical reality, just in the middle, between possibility and reality." So, for Heisenberg and Bohr, multiple possibilities became a single reality at the moment they were observed.

In our large, visible world, we are used to having only one correct answer to each question. That's all Bohr, Planck, Schrödinger, and Einstein wished to achieve: to identify a single, classical deterministic universe from the multitude of quantum uncertainty—a single wave function. Bohr's suggestion that none of the wave functions corresponded to real particles or universes except the one we found when we observed it in some ways managed to evoke a visible world where everything was determined with 100 percent certainty. But it failed in one important way—Bohr's reality didn't give Einstein and Schrödinger what mattered to them the most: an *objective* reality. Another Bohr observer could conceivably observe the same particle (or the same group of particles) and declare a different outcome, also with 100 percent certainty. Different observers could produce different findings for the same particle, and each observer would state with total certainty that his or her finding was the real one. And they would all be right. How is that for subjective reality?

Bohr's collapse of the wave function also committed an additional sin: It created a double standard. It treated the observer (not necessarily a human) as a large, visible entity—in other words, part of the world of classical physics, where there is only one answer to every problem—rather than as another quantum object. In so doing, Bohr's collapse of the wave function mixed determinate, classical physics into the highly indeterminate, quantum world.

To understand why this double standard was such a problem, imagine a situation where all the judges (the observers) ruling on human "quantum particle" disputes and offenses in the courtrooms on planet Earth were imported from an alien civilization on the theory that these alien judges would be independent observers. However, having *aliens* judge *humans* would force the humans to modify their laws and rulings to obey alien law and alien legal rulings. In the same manner, if tiny quantum particles were being judged by large, classical observers, they would have to become and behave like large, classical particles.

Despite the inconsistency in the Copenhagen interpretation of quantum mechanics, the role of the observers in cosmology remains to this day open to fierce debate. But what ultimately undermined Bohr's efforts to create a world run by observers was not this classical/quantum inconsistency. Rather, it was a thought experiment by Erwin Schrödinger, the Austrian physicist whose namesake equation is one of the cornerstones of quantum mechanics.

Schrödinger wanted to extract a single, classical world from the workings of quantum theory as much as Einstein, Bohr, and Planck did. But he resented the godlike status that Bohr gave to his arbitrary observers by granting them the power to decide what was real in the world and what wasn't. Schrödinger corresponded extensively with Einstein over these issues and their paradoxes, and in 1935, he came up with his own thought experiment: Schrödinger's cat experiment. It was designed to highlight the flaws in Bohr's collapse of the wave function, but it has become one of the most famous thought experiments in all of popular culture.

In his thought experiment, Schrödinger imagined a cat locked in a box that also contained a tiny amount of a radioactive substance that might or might not decay after one hour, a hammer, and a flask of poison. If the radioactive substance did decay, it would trigger the hammer to break the flask of poison, which would kill the cat. An hour and a speck later, an observer would open the box and find out if the cat was dead or alive. If the atom had decayed, the cat would be dead. If it had not, then the cat would still be alive. Dead cat and alive cat were both allowed possible states—or wave functions—to describe the cat; each had a 50 percent chance of being correct.

If we relied on the reasoning in Bohr's collapse of the wave function, Schrödinger argued, we wouldn't know before we opened the locked box in which of these two states we would find the cat, and therefore, the cat, before we observed it, had to be in a superposition of both states, meaning it could simultaneously be both dead and alive inside the box. Only the observer who opened the box would know for sure what state the cat was in. In Bohr's terminology, the observer would collapse the superposed wave function by reducing it to only one choice: the cat was either alive, with 100 percent probability, or dead, also with 100 percent probability. So, once observed, the cat could no longer remain in a quantum state where it could be both dead and alive; instead, the cat would be abruptly pinned to one state of being.

Schrödinger was trying to demonstrate how ridiculous it sounded to have an observer pick one solution among the superposition of wave probabilities by offering a paradox: If the cat was found alive, then how could the cat have been either half dead and half alive or completely dead a second before it was observed? The Austrian's thought experiment had the intended effect: Einstein liked Schrödinger's paradox so much that, for a special effect, he suggested that instead of poison, Schrödinger could use gunpowder, which would blow the cat to bits. This would make the paradox even more dramatic: How could the cat inside the box exist in a combined state of being both alive and also blown to bits? At the moment when the cat was found to be alive, would those recently blown-up bits suddenly reassemble themselves into a live cat?

With the unfortunate cat as proof, Einstein and Schrödinger remained adamant in their conviction that there could be only one universe, ours, and that our universe is deterministic—it behaves according to a set of fixed rules (the laws of nature) that exist independently of an observer. If their convictions were correct, then it followed that they could reconstruct the universe's past all the way to its moment of creation and that they could also predict its future. The only thing they had to do was figure out what those fixed rules were!

Einstein and Schrödinger devoted their lives to the search for this rule book, the fundamental theory of the universe: a unified description of all the forces of nature that reinforced order and restored certainty to the universe's past and future, its origin and its destiny. A theory of everything for a single universe would replace the "dangerous" probabilities of the many indeterministic worlds of quantum theory. It was a pressing and enticing goal that seemed within their grasp. After all, in the previous century, Maxwell had achieved the elegant unification of electric and magnetic forces in his theory of electromagnetism. How hard could it be to extend that to the rest of the forces of nature?

As it turned out, finding a theory of everything was harder than anyone could have anticipated. So hard, in fact, that in their race to achieve this goal, Einstein and Schrödinger fell out with each other to the point where Einstein nearly sued Schrödinger for plagiarism—and might well have done so had it not been for Nobel Prize–winning quantum physicist Wolfgang Pauli, who convinced Einstein to drop the fight.

These pioneers of quantum theory failed to find the unified theory of nature or successfully collapse the wave function. But embedded in their failure was a success story: their work built the intellectual foundations on which the search for a testable theory of the multiverse could begin.

I could not have envisioned and developed my own theory without Einstein and Schrödinger's struggles—or without the work of another young physicist who was active around the same time. Yet the story of this intrepid scientist underscored the costs of straying too far from the herd of mainstream theoretical physics.

From 1953 to 1956, Hugh Everett III was a graduate student at Einstein's longtime home of Princeton University. There, the young scientist worked with John Wheeler, a renowned theoretical physicist who was a student, friend, and collaborator of Niels Bohr (not to mention a key figure in the Manhattan Project, which produced the first nuclear bomb near the end of World War II).* Everett found Schrödinger's half-dead, half-alive cat irresistible. It got him thinking hard about the role of the observer and the collapse of the wave function, and this investigation became his PhD dissertation. Everett brilliantly identified the fundamental discrepancy between Bohr's collapse of the wave function and the paradox of Schrödinger's cat: Bohr's hypothetical observer, Everett noted, lived by the rules of a classical world, whereas Schrödinger's cat and the box that imprisoned it were subject to the rules of the quantum world. Everett offered a simple way out of the cat's paradox: Let us agree that everything in the universe, including the universe itself, is governed by one and only one set of rules, those of quantum theory.

The implications of this insight were huge. The observer, like the cat, could now be in a superposition of two states, both dead and alive. Likewise, the cat was now "watching" the observer according to the same laws that governed the observer while he or she had been "watching" the cat. (For all we know, the observer could have been another cat!) Furthermore, the observer's

---

* Wheeler is credited as being the grandfather of general relativity in the United States because he revitalized interest in this field and trained the first generation of general relativists in the country. This core group of experts, known in the scientific community as "Wheeler's kids," went on to train the second generation of relativists, who proudly called themselves "Wheeler's grandkids," and gradually, Wheeler's family grew to become the leading worldwide community in the field of relativity. In 1935, Wheeler was hired as an associate professor in the physics department at UNC Chapel Hill.

states needed to be combined with the dead-or-alive states of the cat to form a single wave function. This meant that there could be a universe in which both cat and observer were alive, another universe in which the cat was alive but the observer was dead (and vice versa), and another one in which both cat and observer were dead. Also, the observer and cat were interacting by the very act of watching each other. This interaction of the cat with the observer allowed them to "communicate" their findings instantly (recall how our tiny, quantum, inflating universe communicates during cosmic inflation) and make adjustments accordingly, thereby giving rise to yet more possible worlds. Mind-bogglingly, quantum mechanics allowed for all of these universes to exist simultaneously.

After Everett's theory, the short version of which was published as his PhD dissertation in 1956, certainty about the world was gone. If Everett was right, then not only the cat but also the observer was a quantum object. It meant the observer had the same status as the cat and played by the same rules. Moving from the realm of exploding cats to the entire universe, Everett showed that a straightforward application of quantum mechanics to the universe predicted a complex and bizarre cosmos of multiple worlds, intricately entangled and wrapped—or superposed (recall the double-slit experiment and our concert hall and the superposition of waves)—with one another into a universal wave function. If, in its infancy, the universe was of subatomic size or nothing more than a tiny particle, then, Everett reasoned, the whole universe, regardless of size, must be subject to the rules of quantum mechanics. Like any quantum particle, it can be represented by a wave function or, more accurately, a bundle of waves packed together into a wave function of the universe.

Just as quantum particles had a chance to take any of a number of trajectories, not just one fixed, predetermined trajectory, likewise (according to Everett) the wave function of the universe has no predetermined course. It can split continuously into many possible trajectories or branches, each of which can produce a different universe. And just like that, the possibility of many worlds returned, over two thousand years after it was considered by the ancient Greeks as a philosophical thought experiment. But which one of those worlds is real?

The answer, according to Everett, is that all possible quantum universes that are superposed into a grand wave function of the universe—that is, all the branches of the wave function that produces universes—have equal chances to exist. They have that right of existence simply because, as Everett argued, no law of nature existed to tell us otherwise. Everett named his theory "the universal wave function of the universe."

The collection of Everett's universes became known in popular culture as "parallel universes." Every time you had to meet a deadline and worked late, there was an identical copy of you at home in another universe reading a bedtime story to your child instead of working. Every time you tweeted something that you wished you hadn't, there was an identical copy of you in a parallel universe that decided not to. Every time you hesitated and weighed your decisions, copies of you in parallel universes were experiencing all the possible decisions that you could have made but didn't in your current universe. The universal wave function of the universe kept splitting and branching out in maddening ways into many possible worlds that accommodated all possible events that could happen to any particle in the universe. As had happened so often before in the history of quantum theory, what was intended as an investigative mission to demystify the weirdness of the theory actually made it stronger and weirder.

Alas, Everett's career was eventually ruined because he went against mainstream thinking. What was Wheeler, Everett's supervisor, to do with these ideas? Bohr was his friend and his hero. Everett was his student. Wheeler couldn't abandon either. Despite Wheeler's penchant for exotic solutions to hard problems, despite his uncanny intuition for recognizing great ideas such as the one advanced by his student, and despite the fact that Wheeler understood Everett was making an excellent point about the faults with Bohr's collapse of the wave function of the universe, it was a very tough choice. Bohr was a force to be reckoned with. Wheeler tried to arrange a meeting between Bohr and Everett so the two could hash out their disagreements. Wheeler asked his star student Charles Misner (who, in addition to being brilliant, was known for a kind and calm disposition) to accompany Everett on their trip to Denmark. (Misner was later my general-relativity professor at Maryland.) However, nothing went according to Wheeler's plan; Bohr and Everett could not be reconciled. Bohr hated Everett's criticism of his treasured, still-dominant theory of the collapse of the wave function. His mind was closed to Everett.

In the end, caught between the outrage of Bohr and his Princeton colleague Einstein, Wheeler decided he couldn't publicly support Everett's many-worlds interpretation of quantum mechanics, so he cut most of the controversy out of Everett's 1956 PhD dissertation by condensing its one-hundred-plus pages down to about thirty. Everett was squeezed out of an academic career and ended up working in the defense industry.

The world would not have known of Everett's many-worlds interpretation had it not been for one man, my colleague in the theoretical physics group at UNC Chapel Hill, Bryce DeWitt. As a scientist, he was as conservative as they come, and as a person, he was every bit as fearful. (He built a bunker in the backyard of his house in Chapel Hill to protect his family from a possible nuclear attack during the Cold War.) DeWitt had spent his professional life working toward the same goal his scientific heroes had—mathematically trying to unify quantum mechanics with gravitational forces. Yet in 1973, as editor of the *Reviews of Modern Physics,* he recognized the importance of Everett's work, took the initiative, and had the courage to publish Everett's complete dissertation.

When Everett's unbowdlerized work was finally published, DeWitt decided to give his "universal wave function of the universe" a better name. He repackaged it as "the many worlds interpretation of quantum mechanics," a name that has stuck to this day. Not that Everett cared. It was too late for him to restart his academic career. He died less than a decade later, in 1982.*

Quantum theory had come full circle, from Planck's quanta of energy to Everett's many worlds, and this evolution had profound implications for cosmology. In quantum theory, the fundamental nature of the world became an unpredictable collection of many possible quantum universes, each of which had a chance to exist. Sometimes they might bundle together in a fuzzy superposition producing many possible combinations. Other times, they might branch off individually. The result: an infinite number of such worlds and an infinite number of possible behaviors. Did our universe, and many others, really result from a game of chance? Until relatively recently, very few scientists were willing to risk their careers on the many-worlds quantum theory. Everett's fate was a cautionary tale for many of us. Including me.

---

* The trauma and struggles that Everett endured while trying to defend his many-worlds thesis and the devastation it brought to his later life are described beautifully in Peter Byrne, *The Many Worlds of Hugh Everett III* (New York: Oxford University Press, 2010).

6

Eleven Dimensions

I N M Y O W N cosmological journey, it helped that my work was solely driven by my love of the subject I was studying. Somewhere along the way, I learned that such an approach, while not always casy or wise, is nevertheless privileged. Being a scientist for the love of science gave me a lifetime of childhood, the kind where you are not afflicted by assumptions or prejudice and where you have no favorite sides to pick, only an exciting new world awaiting your discovery.

When I was a teenager in Albania, there were two mandatory subjects for all university students, including those studying the sciences: first, the history of Marxism, and second, physical education (better known as PE). I abhorred them both, albeit for very different reasons.

I deserved to fail in both subjects. But if I had, I wouldn't have received my degree in a subject I *did* care about: physics. Memorizing the names, dates, and places of Marxist history, especially given that much of what we were told were lies, was torture for me. I've never been good at memorizing. And I have a character flaw—I can happily work around the clock, with the kind of intense concentration like nothing else around me exists, on subjects that I like, but I am a terrible procrastinator on subjects that I don't. So I left the Marxist-history exam preparation for the last day. When I finally forced myself to study, I used a trick I occasionally relied on: I stayed up studying all night before the test. I planned to be the first student to appear in front of the committee the next morning. I would take the exam while everything was still fresh in my mind, then go home, sleep, and forget everything I'd learned. My dad stayed up with me, quizzing me.

But it didn't work. During my exam, as I stood before the committee, they told me they had a very easy question: In what year did Stalin die? I had absolutely no idea. Two of my classmates were waiting outside, and one climbed up onto the other one's shoulders, pressed his face to the window, and held up five fingers, then three. But I learned this later; at the time, I was too embarrassed to look at them as the committee awaited my answer.

After a brief pause, I told the exam committee that I was sorry, but I couldn't remember. Then I burst out laughing and ran out from the room while the committee members shouted after me, "Do you think this is funny? You failed. You don't even know when the leader of the Soviet Union, Comrade Stalin, died?"

Outside, some of my professors had heard the noise and asked what happened. They entered, spoke to the committee, returned, and told me, "The committee has no doubt that of course you knew the answer, but exams can be hard psychologically. They recognized that you just had a temporary blank, and therefore they have decided to pass you." I learned later that my professors had pressured and bargained with the history committee not to fail me, openly hinting that if any of the committee members' relatives took math or physics classes, my professors would go easy on them.

Much the same thing happened with my PE class. I can still hear my PE teacher screaming at me when we were learning how to throw toy grenades and I threw mine right at my feet. "If this were a real grenade, you would have blown yourself and all of us to bits. We would all be killed, thanks to you!" Again, my wonderful math and physics professors had a little chat with the PE teacher, and I passed.

But those experiences made me inherently suspicious of the party line—any party line. It wasn't just what we were forced to learn, it was also what was kept from us. The government felt threatened and could not allow people to peek beyond Albania's borders into the wider world through books, television, and travel. Yet, via homemade antennas and other methods, some Albanians found a way to break those barriers.

Given those experiences, I found it hard to understand the argument that scientists could never explore the moment of creation or what came before it. I couldn't accept this. So, as a young scientist not much older than Everett had been when he decided to buck convention, I chose to explore the confounding world of possibilities that were opening up before me.

By the second half of the twentieth century, the two great pillars of modern physics, quantum mechanics and general relativity, had been chiseled into their current forms. Their establishment, in turn, led to considerable advances in particle physics, which had been around for only several decades but which was poised to revolutionize the world of physics.

Particle physics studies the creation and interactions of sub-atomic species, known as elementary particles, in the universe.

Nested in the theory of the universe is a theory of quantum forces and their particles that populate the universe: a standard model of particles. Particles are carriers of forces.* Throughout the 1970s and 1980s, particle physics seemed poised to unify the quantum forces into one theory. The spectacular strides in particle physics culminated in the unification of the three quantum forces into models of a grand unified theory. Not a small feat!

The unification of forces under one umbrella theory makes use of the fact that the strength of forces (determined by empirical constants, known as "coupling constants") is not really constant; rather, they change (or, in physics terms, "run") with energy. Their unification happens because at some energy scale—in this case, at only about ten thousand times less than the energy of the Big Bang—the forces become roughly of equal strength and thus are indistinguishable from one another.

The three unified forces, out of the four known to exist in our universe, are Maxwell's electromagnetic force, which describes light and electromagnetic interactions; the weak force, which is responsible for particle and radioactive decay; and the strong nuclear force, which binds quarks together to make protons, neutrons, and atomic nuclei. The unification of the fourth known force, the gravitational force, with the other three was the final barrier left between science and the discovery of a mega-theory. How hard could that be?

---

* Each force is achieved by exchanging or scattering its respective bosons—particles with integral spin—to transfer quanta of energy that obeys its own statistical rules. For example, we can think of the gravitational force as an exchange of gravitons, the electromagnetic force as an exchange of photons, the strong nuclear force as an exchange of gluons, and so on. In contrast, half-spin particles are called fermions and they obey different statistical rules.

It couldn't be that hard for the brightest minds. Hawking, who, along with many others, spent a good part of his life working toward the theory of everything, had predicted that it would be discovered before the year 2000.

But yet again, nature refused to cooperate and reveal its theory of everything. Quantum theory *and* Einstein's theory of relativity still could not be unified; they remained locked in the same battle of dueling answers for the workings of the universe that had raged since their respective discoveries.

So theoretical physicists did what they have done for years; they went back to the mental drawing board and began envisioning not a bigger world but an even smaller one. The result, called string theory, was developed by a number of prominent physicists over several decades in the second half of the twentieth century. In the simplest terms, string theory reduces the world to one-dimensional objects, replacing point-like particles with extended one-dimensional strings. There are two possible kinds of strings: open strings, where the two ends are free, and closed strings, where the two ends come togther in a loop. They are too small to be observed. But, string theorists argue, these strings are the essential building blocks of the universe. They are what ultimately made all the elementary particles in the universe; they are the threads from which the very fabric of space-time is woven.

String theory was intended to be the elusive theory of every-thing that connected the three quantum forces to the fourth, gravitational force. But in string theory, achieving the unification of forces in a mathematially consistent way is possible only if the world is enlarged by introducing additional spatial dimensions to the universe. String theorists postulated that, in much the same way as the vibrations made by an instrument's strings give rise to musical notes, the vibrations of tiny string filaments gave rise to different particles; each elementary particle was determined by the subtle intonations of its underlying closed strings. Thus, every particle that had been discovered could be reconceptualized as a "note" in nature's grand symphony.

We are used to thinking of all elementary particles, such as the ones that make up an atom (electrons, protons, neutrons), as point-like particles. But if we were to look at these objects under a microscope with sufficient magnification to observe individual atoms, instead of a particle, we would see a bundle of waves tightly packed together into a tiny wave packet (as you might expect, if you recall the concept of wave-particle duality from our tour of the quantum world).

But suppose we get our hands on an even more powerful microscope, one that can probe sizes much smaller than the size of an atom—one that allows us to zoom in at scales of 10^{-33} centimeters, or a Planck length, the smallest scale at which our quantum and relativistic theories can still be trusted. At this level, instead of glimpsing point-like particles and bundles of waves, we should see loops of vibrating strings. Furthermore, each frequency of each string should correspond to a certain amount of energy. (Einstein's famous equation $E = mc^2$ means that the Planck energy $E = h\nu$ produced by the vibrations' frequency is converted into the mass of the particle.) When vibrating at one frequency, a one-dimensional closed string would correspond to an electron, but vibrating at another (higher) frequency, that same one-dimensional string would produce a proton, and vibrating in yet another frequency, it would produce a graviton (a hypothetical particle that mediates the force of gravity). That is, the type of vibrations of a single string are what determines the mass of each type of elementary particle.

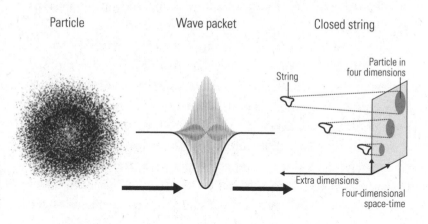

Figure 8. A point particle, in the left panel, is spread into a bundle of waves packed together, shown in the wave packet in the middle panel. But what appears as a point particle to us is, according to string theory, actually the vibration of a closed string, shown in the right panel. The frequency of vibration determines the mass of the particle.

In string theory, the layers of vibrations of the strings—like the notes played simultaneously by different instruments in an orchestra, to stay with our music analogy—combine in a variety of ways to fill every cell of the cosmos, all the way down to the smallest scales. But we can produce the music of the spheres only if we can make this framework harmoniously hang together mathematically. The consistency in that celestial melody comes at a heavy price. Because string theory needs to be reduced to our four-dimensional world.

We humans experience the volume of space in three dimensions: height, width, and length. With the addition of time—which Einstein placed on an equal footing with the other dimensions in his general relativity theory—we have a total of four dimensions (space-time).

Human perception can accommodate a four-dimensional reality. But string theory requires us to assume a world composed of *eleven* dimensions. This mind-boggling idea originated with Edward Witten, a pioneer of string theory and a renowned physicist and mathematician at the Institute for Advanced Study at Princeton, who in 1995 realized that previous versions of string theory could be unified under an umbrella theory he named M-theory. M-theory tells us that besides the familiar four dimensions of time, width, height, and length, there are an additional *seven* dimensions hidden in the cosmos.

Of course, the circuitry of our brains is not wired to comprehend seven extra dimensions. So we have to reframe the concept in our minds. As the great surrealist painter René Magritte, perhaps best known for his self-portrait in a bowler hat with a green apple representing his face, once said, "Everything we see hides another thing." With this statement as our guide, we can start to wrap our human minds around the eleven dimensions that string theory requires.

In an effort to make eleven-dimensional space-time understandable, let's try our own thought experiment using an analogy with paintings and artists. Artists are able to construct a three-dimensional representation of the world on a two-dimensional plane, the canvas, through the clever use of focal points and perspective. Borrowing from their methods, we can visualize M-theory's additional spatial dimensions.

Pick your favorite representational painting that captures three dimensions and look closely at the distance between two objects that appear to be situated at different locations. If you were to shine a light on each object, the light rays are the perspective, and the point where these light rays converge is the focal or vanishing point. Note how the vanishing points of the two objects in the painting are misaligned to be out of focus with each other. This subtle misalignment in the painting is what renders the third dimension possible, because our perception interprets it as depth. While it is impossible for depth to be included on a two-dimensional canvas, the artist has nevertheless managed to draw it by visually tricking our brains.

Exactly the same trick can be used to transcend the three-dimensional canvas of our minds to envision additional dimensions. Although we know that technically a macroscopic fourth spatial dimension does not exist in our universe, we can imagine one.

To grasp how this might work, hold a sheet of paper in a vertical position and at a right angle to your body, as depicted in figure 9. All you should be able to see is a line, not the plane of paper, because from this angle, the length of the paper is at 90 degrees and cannot be seen, and the width of that piece of paper is too small to observe from far away, so therefore it is also hidden from view. All we can see is the height of the piece of paper. For the sake of argument, let's assume that there are more dimensions in addition to the known three dimensions of our universe hiding behind the width and the length of this piece of paper (which, although they are hidden from view, we know exist, because we know we live in a universe with three spatial dimensions). Imagine a fourth and—why not?—a fifth spatial dimension are nested within the width or length of the paper and hidden from view, as in figure 9. These additional dimensions are so microscopically small and curled up that we cannot see them with the naked eye or with our current microscopes, and yet we suspect mathematically that they are there.

To imagine a sixth dimension, suppose (hypothetically) that what is hidden behind the fourth dimension of figure 9 is an object with three-dimensional volume instead of the two-dimensional piece of paper in our illustration. In this case, we would have two more dimensions hiding behind the fourth one, which brings us to a total of six dimensions.

We can repeat this mental exercise to imagine still more hidden dimensions until we eventually get to the eleven dimensions of the M-theory space-time. Thus, much like uncovering tinier and tinier versions of nesting Russian matryoshka dolls, we can extrapolate the other seven extra dimensions of M-theory curled up and hidden from view inside the cosmos by continuing to delve deeper into the layers of space-time.

In sum, the basic concepts of string theory are that point particles hide a vibrating, closed string at Planck length, and space-time hides an additional seven dimensions—but only, crucially, at the kind of subatomic scales that escape testability.

String theory is an interesting exercise—but it is a mathematical exercise. How can we possibly know if it physically exists? String theory can achieve this goal only if it ultimately reproduces and explains the universe in which we exist.

To reproduce a four-dimensional universe out of an eleven-dimensional world, string theorists concentrated their efforts on minimizing or simplifying the seven extra dimensions to return M-theory to a four-dimensional universe. They did so by conceptualizing the added seven dimensions as infinitesimally tiny while keeping the three dimensions of our known world large. This made the extra seven dimensions, though still part of the fundamental building blocks of nature, completely invisible to "normal" creatures familiar with a classical world only, one containing a large three-dimensional volume of space and a fourth dimension of time. Put differently, after having struggled to visualize an additional seven dimensions of space, you are about to scrap them to recover your real-life, four-dimensional world.

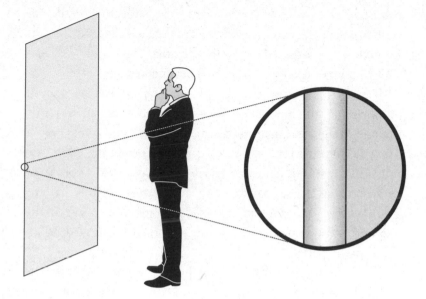

Figure 9. Cartoon illustration of the perspective explained in the adjacent text. The person standing in front the paper can see its length and height but not its depth, and thus he is not aware of the existence of this third dimension. This person would miss out on the existence of additional dimensions and structures if more of these were nested and hidden from view within the depth of the paper.

In mathematical terminology, this curling up of the extra dimensions is known as the compactification of space. But implementing compactification is challenging. If, at the fundamental level, nature has ten spatial dimensions and one time dimension, then what happens to all the other "stuff," such as particles, quantum fields, currents, fluxes, and forces that must fill the extra space in those dimensions, just as particles, quantum fields, currents, fluxes, and forces fill the volume of our four-dimensional world?

Compactifying the extra dimensions is akin to taking a cube full of stuff and trying to squash it down all the way until it completely flattens to its base. Now imagine that cube is filled with quantum stuff, like the fields and the particles and the fluxes and the currents. Every time it is squashed down, the quantum stuff inside is excited, and the interior energy of the cube is sedimented at its base.

These variations on the energy contents inside the cube in our example, the fluctuations from compactification, are nothing more than the quantum fluctuations we came across before. They are best pictured as invisible springs inside the cube, very much like a spring mattress. Suppose a large person volunteers to lie on top of the mattress and compress it until it flattens. Then, as the quantum contents of the cube—the springs in our metaphor—are squashed down, they get further engaged or "excited" and strained.

To add to the difficulty, imagine the height of the cube is seven-dimensional and the base of the cube is three-dimensional. As we squash the cube, all the quantum material inside its volume is compressed and excited along the seven-dimensional "height." This heap of energy has to be dumped somewhere when the cube is flattened, and that somewhere can only be on the remaining three-dimensional surface of the base.

At this stage, if we could zoom in using our super-microscope again, we would expect a glimpse of the hidden world like the one in figure 10. Pictorially, that is how, after the massive mathematical exercise of compactification, string theorists believe they have created a four-dimensional world (as in our universe) that contains matter and energy.

The hope was that once this exercise was successfully completed, string theorists would have derived a unified theory of everything for our origin and would thus be able to write the final chapter in physics.

But this is not what happened. What actually occurred around 2004 was a lot better—or a lot worse, depending on whom you ask.

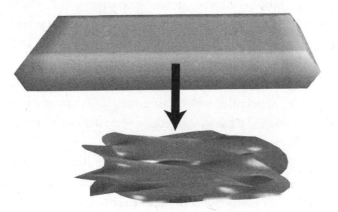

Figure 10. Shown in the picture at the bottom are bundles of strings inside the four-dimensional "compressed spring box," which is what is left after the additional seven dimensions are compactified. According to string theory, pictorially, this is what the space-time of our universe would look like if we could probe scales of, say, 10^{-30} cm. Instead of seeing point particles or empty space-time, we would observe bundles of strings. The volume of the seven additional dimensions of the box at the top is compressed (represented by the arrow) and hidden from view.

· · ·

At first, the outcome of compactification struck the scientific community as a disaster.

Compactifying, a process to get rid of the extra seven dimensions and reduce the volume of space-time from eleven dimensions to four, indeed produces a universe like ours. But it also gave physicists far more than they had wished for.

It turns out there are a great many ways of curling up the extra dimensions and even more ways of combining layers of fluctuations of the quantum stuff in that extra volume. And for each possible option, another potential energy well,* known as landscape vacua, that can potentially ignite a Big Bang is produced. At present, using the mathematical process of compactification, string theorists have found approximately 10^600 (10 with 600 zeros behind it) possibilities. This vast collection of potential Big Bang energies, the collection of about 10^600 vacua obtained through the process I have just described, is known as the land-scape of string theory.

The discovery of the landscape of string theory shook the world of theoretical physics to its core. The decades of efforts by string theorists to mathematically reduce the eleven-dimensional world of string theory to obtain solutions that describe a single four-dimensional universe had inadvertently unleashed the scenario of a virtual universe-making factory that could act as an incubator for many potential Big Bang energies from which billions of baby universes could possibly spring into existence.

If that seems like a lot to absorb, it is—even to seasoned physicists. However, there is a way we can envision a string-theory landscape as somewhat like the physical landscapes we are familiar with. It can have peaks and valleys, like a mountain range. But unlike the regular landscape on Earth, which exists in real space and time, the string-theory landscape exists in a space of energy—energies that represent the range of choices or possibilities that exist in this string world when trying to produce a four-dimensional universe. Just as on the familiar physical landscape a handful of marbles might roll down from a mountaintop and settle in a valley, in the string-theory land-scape, a whole infant universe could settle in a place, a vacuum site. But the string-theory landscape contains a vast number of

---

* Drawing from our analogy with the marbles rolling down a mountain range, I will sometimes refer to these vacua as energy valleys instead of energy wells.

choices of vacuum energy sites from which our universe could have started. The new mystery in the context of the landscape discovery is which vacuum energy did our universe pick, and why did it pick that one?

The discovery of a string-theory landscape meant that, instead of *one* initial energy for *one* Big Bang type of inflation that resulted in *one* singular universe, there were an abundance of energies— many potential Big Bangs to start multiple, four-dimensional universes like ours. Instead of once and for all explaining our origin, this abundance of potential energies, trillions and trillions of possible starting energies so far (as illustrated earlier in this chapter), offered a mystifying multitude of possible origins. Simply put, a string-theory landscape provides a vast collection of initial energies—of potential Big Bang energies—capable of jump-starting multiple universes.

Unexpectedly, at the beginning of the twenty-first century, string theory had dealt a major blow to the simple vision of a single universe wrapped up in a theory of everything. By predicting a landscape of many possible worlds like our universe, string theory unintentionally traded the theory of everything for a theory of the multiverse, precipitating a major crisis in physics. For researchers around the world, this discovery seriously threatened the possibility that string theory would become the long-sought theory of everything.

Only a few years previously, in his book *A Brief History of Time,* Stephen Hawking had summarized the desire and the millennial efforts for the dream of a single universe described by the theory of everything and declared his advocacy for the theory of everything by paraphrasing Saint Augustine: "When asked: 'What did God do before he created the universe?' Augustine didn't reply: 'He was preparing Hell for people who asked such questions.' Instead, he said that time was a property of the universe that God created, and that time did not exist before the beginning of the universe." Still, while scientists were searching for a theory of everything and a singular universe, the landscape discovery implied that the world of physics was about to stumble away from its goal and into a terrifying inferno.

Physics was preparing for a paradigm shift, where every dream of a single universe wrapped in a theory of everything was about to be shattered into smithereens. It was far worse than the previous clashes among the quantum theorists, such as the fierce arguments between Einstein and Bohr over a deterministic classical universe versus an indeterminate quantum one; once, when a frustrated Einstein quipped that "God did not play dice when making the universe," Bohr retorted, "Einstein, stop telling God what to do."

Backing down was not an option unless the whole endeavor of string theory was to be discarded. Again and again in our theories of nature, each time we asked: Where do we come from? The answer was: A multiverse. Understandably, to most people working in physics, the premise of a string-theory landscape was nothing less than a full-blown crisis.

Indeed, the string-theory landscape did not come with a manual that explained why some of these possible universes might be better or fitter than others. It appeared that each one of the potential four-dimensional worlds out of the pool of 10^{600} was an equally likely candidate to start our universe. It was like Schrödinger's cat and the Hugh Everett paradox all over again. Moreover, since, bound by the speed-of-light limit, we could not observe beyond our universe's own horizon, testing for and proving this new multiverse possibility seemed hopeless.

There was something else that made the multiverse scary for scientists: Hugh Everett's fate. His example provided a poignant reminder of the risk of siding with a multiverse scenario of the cosmos.

To the majority of scientists in the theoretical physics community, the string-theory landscape seemed to be the worst of all worlds.

But not to me.

For me, the string-theory landscape arrived at exactly the moment when all my thought experiments about different starting points for different universes had been completely undone by the rules of entropy and the second law of thermodynamics. The discovery of the string-theory landscape was the third time that a physics breakthrough produced the same answer to the question of where we came from.

To my mind, the possibility of a multiverse could no longer be ignored.

7

First Wave

THE DISCOVERY OF the string-theory landscape serendipitously came to light just as I started researching the creation of the universe full time. When I arrived at the University of North Carolina at Chapel Hill in January 2004, I was convinced that to meaningfully inquire about the origins of our universe, I needed a pool of infant universes from which to choose. However, I had no preconceived notion of what the ultimate theory of the universe should look like, so I began my investigation of the universe's origins with no idea of where it would lead.

I did, however, have a sense of where it would probably *not* lead—to an increase of my chances of tenure. Under the pressure of tenure decisions, assistant professors work long hours, and most decide to join established fields of research, such as, in my field, post-inflationary cosmology. With my best interests at heart, my mentors had advised me to follow this path and avoid working on high-risk challenging problems like the creation of the universe, at least until I had achieved tenure. They recommended that I work on more conventional topics and join large research groups led by well-known authorities.

Given that only about 30 percent of PhD graduates are able to secure jobs in academia and fewer still are able to advance on the tenure track, this advice was intended to help me thrive in my new profession and make sure I stayed in it. Some of my peers in the broader physics community were clearly following similar advice; they spoke of being careful to map their career paths long in advance by joining the "right" camp while not upsetting anyone in the "rival" camp. I heard tales of the consequences of going against the trend, but even so, I could not fully comprehend the concern that openly debating ideas with other scientists could be considered personal attacks against them—and indeed, once I found my niche within the world of physics, this proved not to be the case.

To be sure, I was fully aware that working on risky, envelope-pushing subjects was unwise from a professional standpoint; I only had to look at the sad example of Hugh Everett to understand that. But I had become a theoretical physicist precisely because I was fascinated by problems such as the beginning and the end of the universe, and I had already devoted all my spare time to these subjects before arriving at UNC. I found it very hard to resist the attraction of what had gotten me interested in science in the first place.

When in doubt on a major decision, I have always picked the choice I won't regret later. And by the time I arrived at UNC Chapel Hill, I knew that if I didn't work on the universe's origins, I might regret it for the rest of my life. So that is the field of research that I picked.

Rationalizing my very impractical selection, I told myself that if I had been a practical person, I would have chosen to work in a more lucrative profession to begin with; I would never have picked the path of a physics professor. Then, too, in light of my childhood experiences in Albania, the decision to research the creation of the universe didn't feel like a particularly courageous act. As far as I knew, in modern times in the West, no heads rolled for proposing new ideas, and everyone was free to speak his or her mind—even if doing so ruffled some other people's feathers.

In large part because of my background, I felt protective of the freedom to think for myself. This freedom also seemed like a cornerstone of science; looked at in certain lights, after all, the whole history of science is one continuous, never-ending battle of challenging and improving on ideas that previously were worshipped as "conventional wisdom" or the "final truth" on a subject. And, just like the magic that animates Beethoven's music even hundreds of years after he composed it, good, original scientific ideas (which are few and far between) in the end survive the most challenging test of all: the test of time.

I had prepared myself for the likelihood that I would fail. In fact, the entire field of theoretical physics prepares you to cope with disappointments and failure. For theoretical physicists, a best-case scenario is one where *only* nine out of ten of your ideas are wrong—and even then, most of us never know that we were correct one-tenth of the time, because opportunities for theoretical physicists to test their new ideas observationally are rare.

But where observations fail, the scrutiny of peers comes to the rescue. The theoretical physics community operates like an extended family. The bond among its members is based not on blood but on a deep respect for one another's views. Of course, as in any family, respect has to be earned the hard way—in our case, by contributing to groundbreaking ideas and advancing knowledge. To that end, we scrutinize, criticize, and work hard to pinpoint logical flaws in the ideas of our colleagues as well as in our own. Even if we rip apart each other's reasoning, we remain united by our shared pursuit of the same goal: to learn the true answer to the mysteries of nature.

• • •

As I got to know the culture of my new academic family, I came to realize that, to pursue my investigation of the universe's origins thoroughly, I had to first give serious, sustained consideration to the single-universe view and fully understand the reasons why the scientific community had largely rejected the idea of the multiverse.

I kept revisiting the arguments for a single universe, sometimes during walks on the baking-hot trails and streets of Chapel Hill, North Carolina, accompanied by snakes, deer, beautiful birds, and exotic creepy-crawlies that, before arriving at UNC, I had seen only in the movies. (I was one of the very few people walking to and from work. Locals would often stop their cars and kindly offer me a ride. They must have thought it strange that I was determined to walk to the university, regardless of the weather. But once they learned I was a physicist, they seemed to accept these walks as "normal" behavior.)

As I already knew, and as we have seen, the deeply held belief in a single universe ruled by a single theory was powerfully embedded in tradition. But there was more to its appeal than that. The single-universe view dominated contemporary thinking and a unified theory was the holy grail of physics because scientists, including me, wanted simplicity and testability in the theories that we constructed to explain the universe.

To appreciate the sway that these values hold, consider cosmic inflation. Relying on only one assumption, the inflaton potential energy, cosmic inflation elegantly accounted for the main features observed in our universe; equally important, its predictions about flatness, homogeneity, and uniformity of structures in the universe are testable.

By contrast, the study of the multiverse had remained almost taboo because scientists were convinced that the multiverse could not be tested or independently observed. Einstein's theory of gravitation was based on the postulate that nothing, absolutely nothing, could travel faster than light in the universe. We observe objects—for example, stars—by sending and receiving light signals. The horizon of our universe, which is only about 10^{27} centimeters from Earth (10 with 27 zeros behind it), is defined as the farthest distance that light in the universe can travel and still return to us. The speed-of-light limit will not permit us to exchange light signals outside the horizon of our universe into the multiverse to observe what lies beyond its "edge." If the existence of the multiverse could not be tested or observed, scientists believed, then it could not be a scientific theory. This, I fully appreciated. However, the single-universe scenario suffered from its own share of unsurmountable problems, many of them related to the second of law of thermodynamics, as I described in earlier chapters.

Consequently, I was among the very few physicists to be delighted when, in 2004, around the time of my arrival at UNC Chapel Hill, the discovery of the landscape of string theory took off. In contrast to string theorists, who were working to mathematically reduce or compactify the extra dimensions in order to produce a single universe, I was approaching this problem as a cosmologist. In searching for an answer to two very specific questions—what jump-started the universe, and what was there at a time before our time?—I had convinced myself through various thought experiments that this mystery would make sense only if we had a *pool* of possible origins from which to choose, along with the initial energies with which to inflate them. This pool of possible origins is known in cosmology as "the space of initial states for the universe," and until the discovery of the string-theory landscape, it had been an abstract hypothetical space of possibilities.

Up until that moment, the pool of energies that I was using was merely an abstract possibility. Serendipitously, the discovery of the string-theory landscape offered exactly the pool of initial states of the universe that I needed: a collection of actual energies derived by an underlying theory.

In my view, the landscape discovery was not a disaster; far from it. I thought it was the best possible news for any theory that attempted to explain the origin of our universe and what was there before. With this breakthrough, I could now calculate the probability of our origins using the landscape of string theory as the physical space of potential initial states for our universe. And this derivation of an answer led me to a multiverse.

The discovery of the string-theory landscape had opened up new intellectual horizons for me. But a formidable challenge remained: How could I relate a physical object—a thing that exists in real space-time, like our universe—to the string-theory landscape, an abstract space of energies obtained from an eleven-dimensional world? At first glance, the two seemed incompatible. But I suspected that the underlying connection between our universe and the landscape resided somewhere in the intersection between string theory and quantum theory.

Then, one day, the pieces of the puzzle began to fit together.

When the idea came to me, I was sitting in a Chapel Hill coffee shop. I like working in cafés; like my long walks, sitting in coffee shops gives me the many hours of unencumbered concentration and the thinking time I need to detach from my surroundings and be fully absorbed with the problem in my head. Wherever I live, baristas soon get to know my habits and my love of long espressos, and the anonymous collective buzz of a café is less of a distraction to me than a silent office with frequent knocks on the door from colleagues and students.

I was staring out the window, going through the arguments in my head, when a notion crept in: *quantum mechanics on the landscape of string theory.*

Yes, of course! I thought. Put the two truths together to produce a single, greater prospect.

I started with the idea of our infant universe as a tiny, particle-type object. This meant that I could apply quantum mechanics to it. As I hinted before, thanks to the wave-particle duality of quantum mechanics (which we saw in detail in chapter 3), I could think of our infant universe not only as a quantum particle but also as a bundle of waves tightly packed together into a wave packet to resemble a particle. These wave packets are the branches of the wave function of the universe we came across previously.

At this stage, the connection between what appeared to be two incompatible subjects—a physical universe existing in a real space and time, and a pool of energies existing in a string-theory landscape—became clear. *Let this wave function of the universe pass through the chain of landscape energies,* I thought. *Find out how and where these wave-universes (1) take their energy from the landscape, (2) go through their individual Big Bangs, and thus (3) transition to growing physical universes in space-time.*

It was a simple idea: By expressing the infant universe as a wave packet that travels through the landscape of string theory, I could use quantum equations to find out which energy site it would settle into in the midst of that vastness. Furthermore, the family of solutions, derived from the Schrödinger-like equation, for all the branches of the wave function of the universe provided direct information on their probability of existence. In quantum mechanics, the probability of each wave packet (or branch) solution of the wave function is directly proportional to the square of the respective solution. It was essentially a mathematical "experiment," just as one might set up an experiment to find out what happens to a handful of marbles if we let them roll down mountainous terrain. The energy site that I was seeking was the valley where my quantum-particle marble would end up. Ultimately these solutions to the wave function of the universe and the vacuum energies into which its branches settled would allow me to calculate the probability that the wave function of the universe has to select an initial energy like the one that started our universe.

I wrote the words *QM on the landscape* on a slip of paper as a reminder—not that there was any danger I might forget. Then I decided to call it a day. Putting a set of scribbled ideas away for a while helped me detach myself from them. I would return to them later with fresh eyes and a fresh mind and be better able to spot any blunders or flaws in my logic or calculations.

Before I packed up my notes and headed home, I went outside for a cigarette. During periods of intense concentration, in moments of frustration or anticipation, when I am close to solving a problem but am not there yet, I take short breaks in the fresh air. They help me to be cautious and curb the excitement that comes with a new idea, allowing me to clear my head and put the pieces of an idea together in a meaningful way. Cigarettes, ironically, were my excuse to get some fresh air. I had started smoking years before— the night in Tirana, in fact, when I watched my classmates and friends scale foreign embassy walls to escape Albania. I never saw or heard from them again. Nor did I manage to quit smoking until quite recently. In Chapel Hill, cigarettes and classical music were for many years my daily companions.

As I stood outside trying to visualize the idea, my internal monologue kept going nonstop. *This is all so beautiful,* I thought, *and so simple. What took me so long to see it?*

No—it can't be this straightforward! What am I missing?

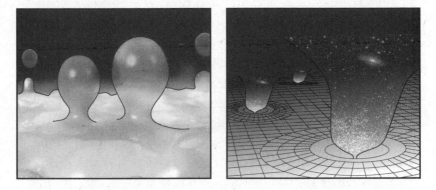

Figure 11. A visualization of the transition from wave-universes settling on the energy vacua of the landscape to inflating physical universes that exist in real space-time.

It seems so obvious. Other people must surely have thought of it and decided it doesn't work.

It must be wrong . . . or dumb . . . or both. But why can't I see what's wrong with it?

These were the thoughts that raced through my mind as I stood in front of that coffee shop.

When I went back inside and packed up my notebooks, I couldn't resist taking another peek at the notes I'd written on the scrap of paper. All the equations suddenly started to flow. I could now see how this idea would take me to the answer to the mystery of the universe's birth. It would allow me to derive the probability of the origin of our universe and thus evade the swamp to which Penrose had banished our chance of existence.

The landscape was exactly the starting point that I needed. And by using quantum rules to calculate the answer instead of just postulating it, I finally had a way to connect the landscape of string theory to our universe's origin.

Importantly, the conceptual step of a wave function of the universe on the landscape energies meant I could merge the two parallel tracks of quantum theory with gravitational theory, since our universe was a quantum wave filled with energy until 13.8 billion years ago, when it exploded into a physical universe governed by this energy's powerful gravitational forces.

Tired and excited, I left for home. It was late, but I had to tell someone what I had found. So I phoned my two biggest fans: my husband (he worked overseas and we were in different time zones) and then my dad. Both knew what I was working on, and I knew they would not mind listening to this new development no matter what time it was.

I explained my idea animatedly to my husband first. And then I called my parents. "Dad," I said when he picked up, "do you remember when we would stay up at night and listen to classical music on Radio Tirana?"

"Yes, of course I remember," he said.

"Dad, I think I may have cracked the problem."

"How?" my dad replied. I could hear Mom protesting on the other end of the line: "I want to talk to my daughter. She's my daughter too." My dad said, "Talk to your mom first," and I knew he was right. My mother had been diagnosed with cancer a year before and had just finished chemotherapy, which had successfully arrested the cancer. But our whole family was extra-gentle around her. She managed to retrieve the phone from my father. "Hello, how are you, darling?" But before she could continue, I said, "Mom, I'm so sorry, but please, this is important, and I haven't finished. Can you put Dad back on the phone?"

I blurted the whole thing out in a few minutes without taking a breath. My dad said, "Hmm," which I knew meant he had concentrated on every word and was thinking. Then he said, "Talk to your mom. She has missed you." This time, desperate,

I protested. "But Dad, first tell me, what do you think? Do you think it is total nonsense?"

"No . . . beautiful," he said. Then both of them said in unison: "When are you coming to visit?" The longing in their voices lingered in my head until I reached home.

I entered the house and headed straight for the bedroom, and I was switching the lights on when I heard knocking at the front door.

I was not expecting anyone. The knocking intensified, and I got a bit scared. A colleague at work had been harassing and threatening me daily—something that is, unfortunately, a routine experience for women who have chosen a career in the hard sciences. I had ignored all his threats and advances, writing them off as a nuisance, which had led this character to nickname me "Her Majesty."

The knocking at the door became more persistent and would not stop. Then I saw a silhouette on the veranda and heard still louder knocks on the veranda's glass door.

I dialed 911. A few minutes later, the police arrived, and I opened the door.

The officer stood there holding a beautiful big bouquet of exotic flowers. "These are for you," he said. "The man knocking on your door and windows was trying to finish his last flower delivery of the day." I thanked him and apologized, and he told me, "Better safe than sorry—you did the right thing."

The flowers were from my husband. I started laughing. Somehow, I had managed to involve the local Chapel Hill police in a weighty academic problem.

8

Into the Multiverse

Practicing science without mathematics is inconceivable to me. Mathematics is the unifying language of the universe; it is the language in which the laws of nature are written—or, in Galileo's words, "Mathematics is the language in which God has written the universe."

I'm going to spare you the full working through of my math, but I am going to briefly retrace the journey I took as my idea germinated. In the process, I hope to share some of the beauty hidden in that math.

As we saw in the context of the string-theory landscape, in physics, a collection of many possible beginnings is generally known as the "space of initial states." If we are speaking about universes, each of these initial states might produce a universe, although almost certainly not all of them do. In my case, the collection of energies presented in the string-theory landscape offered me the space of initial states on which I could perform my calculations.

I was aware that starting with a pool of multiple beginnings opened the door to the possibility of multiple universes—a multiverse. However, rather than assuming that outcome at the start, I wanted to calculate and compare the individual chances of success for each of these initial states. Perhaps the answer would turn out to be that all of them except one had a zero chance to bring a universe into being. Or perhaps many of them were likely to produce their own universes. But I wouldn't know until I sat down to do the calculations.

If the calculations showed that only one universe came into existence, then I would be happy to concede that we had to find a different way to address the serious problem raised by Penrose's claim that our universe had almost no chance to exist. But if my calculations showed that many of these initial states were each capable of producing a universe—if it led to a picture of the cosmos where our universe is almost as unremarkable as a grain of sand on a beach—then I and the rest of the physics community had to start to consider our universe as part of a multiverse, with all of the incredible, Everett-style implications that came along with it.

From there, my mind went to the type of mathematics required for a full implementation of my idea and a plan of the next steps I needed to take. The equations I had to solve for the wave function of the universe propagating through the vast string-theory landscape of energies were very complicated. But if somehow I could do this calculation, then estimating the chance for our universe to exist was straightforward: the probabilities for each possible universe starting out on the landscape were simply given by the square of the respective wave-packet solution. Once I knew these solutions, I could compare which infant universes, out of all our choices, were the most likely to have survived creation and "inflated." And perhaps, finally, I could resolve the question of whether the probability that a universe like ours had to emerge out of this structure was higher or not compared to other possible universes.

The next day, I began the calculations. It was challenging; the mathematics involved were highly complex and required significant work. I had to follow the evolution of all the branches of the wave function of the universe as they spread throughout the vast complicated structure of the landscape energies. But the difficulty did not matter; it was too exciting. If it worked, this approach would reveal how the interplay of quantum and gravitational forces could explain (instead of postulate) the origin of our universe—and, quite possibly, its emergence from the multiverse.

Far from reaching the end of the cosmic story, I felt that, in many ways, the cosmological expedition into the workings of the universe had just begun. Science was poised to make another leap, this time beyond the borders of our universe and back in time to the instant of its creation. Waiting there, hiding in the cracks of the foundations of our beliefs, was the ghost of Everett—and, waiting with him, the theory of the multiverse itself.

This scientific leap, as I envisioned it, was captured in that one phrase that came to me at the coffee shop: *quantum mechanics on the landscape of string theory*. And the theory I was building from it would come to be called the theory of the origin of the universe from the quantum landscape multiverse—or, for short, the quantum landscape multiverse. To this day, it is one of my proudest creations.

To show how I ended up with a theory of the quantum landscape multiverse while trying to understand the unlikely origin of our universe, I need to go into more of the science involved in the calculation and the derivation of the answer. For starters, my theory of the origin of the universe from the quantum landscape multiverse relied on treating the entire universe as a quantum wave packet.

We know for a fact that at its earliest moment, our universe was about a few Planck lengths—smaller than the tiniest quantum particle we know. I reasoned that this fact justified my thinking of applying quantum theory to the whole universe in considering our infant universe to be a branch of the wave function of the universe.

If the wave function of the universe is a bundle of waves, meaning it contains individual branches of quantum wave packets, each of which can potentially seed a universe, then it will likely give rise to a number of worlds rather than just one. Based on the wave-particle duality of quantum mechanics, we are allowed to think of these branches of the wave function of the universe as either a bundle of waves *or* a beam of quantum particles.

Next, let's allow the wave function of the universe (with all its branches) to run loose on the landscape of string theory, and see what happens.

So far, the emerging picture of a multiverse from the wave function of the universe seemed close to the spirit of Everett's many-worlds interpretation of quantum mechanics. But not for long. My first attempt at solving the quantum equations that would allow me to find out which energy site our universe chose from the landscape produced a surprising result.

To grasp this situation intuitively, think of the marbles that our gaggle of physicists accidentally sent rolling down the ragged terrain of mountains and valleys in the Rockies (or in the Lake District, depending on your preference). We already know that the marbles rolled until they settled into the lowest possible points in the valleys.

Like the contours of Earth's gravitational potential energy on a mountain range, the string-theory landscape has its own energy valleys—a whole landscape of vacua. The string landscape has billions of energy valleys, vacua whose depths are spread randomly, ranging from low to high energies. You may recognize *vacua* as the plural of *vacuum*. But in the quantum-size string-theory world, especially the compactified eleven-to-four-dimensional string-theory landscape world, a vacuum is not empty. Rather, it is the default or stable state—it is the point where our theoretical marble can stop rolling.

These near-endless chains of vacua are what supply a landscape of potential energies on which the branches of the wave function can settle. So if we continue with our marble analogy, the landscape of energy valleys is like the gravitational potential of a physicist's classical mountainous landscape. The marbles are the branches of wave-universes or, equivalently (using wave-particle duality), a beam of quantum particles in the wave function of the universe trying to travel through this terrain. As these wave-universes move through the landscape, they sample the energies of the various landscape valleys on which they may eventually settle.

Instead of the example of marbles rolling down mountains to settle in a variety of valleys, perhaps a more useful example to illustrate what was missing in my previous thinking—and the idea that came to me in the coffee shop—is the analogy between the wave function of the universe traveling through the string-theory landscape and a quantum system we are more familiar with: a beam of electrons going through a long piece of wire capable of conducting electricity.

Understood from a quantum perspective, a piece of wire is actually a long chain of billions of atoms and fluxes of electrons trying to travel through the chain without being stuck in any atomic site. In this case, the branches of the wave function of the universe are like the beam of electrons trying to travel along the atoms of that wire, with the energy valleys corresponding to the potential electrical energies contained in that chain of individual atoms of the wire. If we have a perfectly clean piece of wire (a perfect conductor) made of identical atoms with identical spacing and with no impurities or air bubbles, then electricity is conducted evenly throughout the wire, which means that none of the electrons in the beam get trapped inside the wire. All the electrons travel successfully from one end to the other without any losses, thus giving us a perfect electric current. (In real life, wires are not perfect. They may contain bits of impurity or air bubbles. Some of the electrons in the beam may get trapped inside the impure chain and never travel all the way through. In this case, the electric current will suffer a loss, since more electrons are going in than are coming out of the wire.)

In our analogy, the chain of atoms in a perfect wire corresponds to a regular or periodic chain of landscape energy vacua, where all of the vacua have the same energy. But such a perfectly ordered landscape of energies is really bad news if we are trying to harvest universes out of them. Here is why: If you compare the electrons going through the wire to the wave function of the universe going through the landscape, then having a perfect conduction of quantum wave packets means they travel all the way through the landscape without ever getting stuck in any of its energy valleys. If none of the wave packets are confined and settle on top of any of the landscape energy valleys from which they can draw sufficient Big Bang initial energy to fire up a universe, then no actual universes will grow. An orderly landscape with a chain of periodic identical vacua is barren—it looks like a flat, uniform desert.

And so, my final step in putting QM on the landscape—just as we did previously when imagining marbles rolling down a mountain and just as physicists do when they study the quantum behavior of electrons going through a piece of wire—was to solve the quantum equation, a Schrödinger-type equation that describes how a wave or a quantum particle moves under the influence of an external force (potential energy) and the probability the particle has of taking a particular path of motion. I could use these equations to find out what happens to the wave function of the universe when it travels along an "impure wire"— the randomly scattered potential energy valleys of the landscape.

Quantum formalism, when applied to the wave function of the universe, is referred to as quantum cosmology. It is an advanced version of regular quantum theory, but it addresses the motion of waves in abstract spaces, like the space of energies of the landscape, instead of the motion of quantum particles through the real space-time composed of length, width, height, and time.

Quantum cosmology offers a set of equations and rules that describes what happens to the wave function of the universe propagating on abstract spaces, such as the space of energies of the landscape, and how a real universe in a physical space-time is spit out from these initial waves and energies. One of its founding fathers is Bryce DeWitt, the same scientist who defended Everett (and who rebranded his clunky-sounding "universal wave function of the universe" as "the many-worlds interpretation of quantum mechanics"). The other is John Wheeler, who mentored Everett. The quantum equation that gives the probabilities for the wave function of the universe is known as the Wheeler-DeWitt equation.

The Wheeler-DeWitt equation in quantum cosmology is the equivalent of the Schrödinger equation in regular quantum mechanics. And by using a wave function of the universe on the landscape, it means that the beam of branches gives rise to a collection of universes in the same manner as in Everett's many-worlds multiverse. In other words, Everett's many-worlds theory is embedded within this theory.

By applying the Wheeler-DeWitt equation and the quantum probability rules to my idea of a wave function of the universe propagating through the string-theory landscape, I could derive how our universe was selected and which energy site on the land-scape it chose for its Big Bang. Or so I thought back then. As it turned out, the math was harder than I anticipated.

Despite how alluring the concept that had come to me in the coffee shop was, the mechanics of the solution were exceedingly complicated. As you can imagine, a landscape with 10^{600} valleys where our universe could potentially settle meant the re-quired math was horrendous. I had a choice: I could make some crude assumptions to simplify the landscape—say, reducing the size of the landscape to contain two energy valleys instead of trillions of them, in which case the equation became manageable and could be solved by hand—or I could attempt to solve the real equation, without simplifications and with all 10^{600} vacua in it, no matter how long that might take.

I suspected that similar equations might have been solved before by condensed-matter physicists, who study materials like the piece of wire with the electrons. So I consulted with my condensed-matter colleagues at UNC and spent the next six months taking a crash course in condensed-matter physics. It helped me understand and identify mathematically similar struc-tures in the string-theory landscape, which exist in condensed matter in the form of exotic materials known as "quantum dots" and "spin glass." Luckily, complicated mathematical methods (like random matrix theory, which might sound familiar to the advanced reader) for solving these types of problems had been developed in great detail by the condensed-matter scientists.

My hunch turned out to be correct. By the end of my detour into condensed-matter physics, I could finally solve the equation for the wave function of the universe on the true, unabridged landscape of string theory. If I had chosen the easier way and simplified the landscape to only two energy valleys, my answer would have been wrong.*

One of the keys to my solution is the fact that energies in the landscape vacua come in all different sizes and distances. The landscape, in other words, looks nothing like a flat, uniform desert; it's full of peaks, valleys, and hills. Moreover, the distribution of all these asymmetrical features is irregular; it is disordered, nothing like our "perfect wire." This disorder turned out to be crucial.

To visualize what happens in a disordered landscape, let's get back to our example of electrons passing through a wire. If instead of the perfect wire, we have a wire made of an insulating material, a material full of impurities and disorder (for example, glass), then the energies in its chain of atoms are also irregular and disordered. Electrons cannot go all the way through the material; they become trapped. That's why such materials are called insulators. In the large, visible world, we know this to be true: If I shoot an electric current into your kitchen window, the beam of electrons gets confined inside the glass and stays there. Once the electrons are trapped—*localized* is a better term for physics—inside the glass, they are these tiny wave packets confined to energy sites located along the individual atoms in the chain.

---

* In parts of this project, I invited an excellent scientist and wonderful collaborator, Archil Kobakhidze, who at the time was my postdoctoral fellow and now is a professor of physics at University of Melbourne, to join forces with me. His scrutiny and hard work were integral to my efforts.

So what happens to our electrons-and-glass example in the microscopic world? If we were to use an atomic microscope to observe the electrons inside the glass, we would see that the electrons' localized behavior is a textbook example of quantum interference. Think back to our double-slit experiment with light and atoms. Like an ocean wave crashing against a rocky shoreline, the electron wave packet "smashes" along the chain of atoms and tries to push through the wire. But instead, it keeps getting scattered and broken into two parts, a reflected and a transmitted wave, at each of the atom's sites. The more scattering sites the electron wave goes through, the more of these reflected and transmitted waves we have. The result is a bunch of quantum waves inside the material, which we know (from chapter 2) will add up and interfere at every single point.

In the case of a disordered chain of atoms, these waves are out of phase with each other due to this disorder. They are "disoriented" from being scattered from irregular atomic sites. Therefore, their interference pattern is destructive (a phenomenon you saw visualized back in figure 4). For our electrons in a particle form traveling along the wire, this means that an electron will be found trapped around some atomic site (where most of its wave is concentrated) but will likely not be found anywhere else in the wire. (If I had oversimplified the landscape to two vacua instead of zillions, then there would have been only two scattered pieces of the wave to add up. The two-vacua kind of landscape would completely lose the complex interference pattern and localization. For this reason, any simplification of the string landscape into a two-vacua object would have produced the wrong physical result.)

In short, like the electron going through an insulating material, the beam of tiny quantum universes trying to travel through the disordered string landscape gets trapped in the various energy valleys inside the landscape. Rather than passing quickly through the energy field, these tiny wave packets resembling quantum universes get stuck. The disordered distribution of vacua energies on the landscape was the key factor that triggered the localization of the branches of the wave function on its vacua. Once these tiny quantum universes "localize" on a particular landscape energy site, they take that vacua's energy, which drives them through the explosion of Big Bang inflation and makes them grow large.

And this was the picture that emerged out of the equations: As the quantum waves try to make it through the landscape, they get trapped in some vacua, take the vacua's energy to go through an inflationary expansion, and create real, macroscopic universes out of these infant quantum universes. But since different quantum wave-universes settle in different vacua, their Big Bang energies are also different, because it all depends on what landscape energy vacua they find themselves in.

There is an easy way this energy differential can be visualized. If you recall, when I talked about destructive and constructive interference of waves, I used the example of a concert hall where the waves from all the instruments in the orchestra add up to amplify in certain seats in the hall (the expensive seats) and cancel each other out in other seats (the cheap seats). Let's think for a moment of the individual seats in the concert hall as being specific energy vacua on the landscape. The interference among the branches of the wave function of the universe looks similar to the interference of the sound waves from the orchestra in the concert hall. Therefore, a quantum wave packet would concentrate mostly around one site and fall off to zero everywhere else, like a concert hall where there is only one expensive seat—one good seat where the music is amplified—and the rest of the seats in the hall are cheap seats, where the music can barely be heard. This single good seat is the vacua on which an infant quantum universe sits. But which "seat" on the landscape would our infant universe have chosen?

The first time I got to the end of my equations and looked at my solution, I felt dumb, because the answer I got was nonsense. Even worse, it took me back full circle to the original problem of our universe's unlikely origins. The solution I found asserted that the most probable universe was the one that had started at the lowest energies on the landscape, which meant that, according to this calculation, a low-energy Big Bang had the highest chance to produce a universe. Translation of this result: our high-energy Big Bang universe once again had the smallest chance of coming into existence!

With hindsight, I know that this nonsense answer should not have been a surprise. I should have expected it before I sat down to calculate. I should have anticipated that quantum particles, like marbles rolling down mountains, will search for the lowest energy valleys to settle in because they are more stable there. I should have had foreseen this before attempting to solve the Wheeler-DeWitt equations. But I didn't.

After the first time I worked through the equations, I spent any free moment going back through a photographic copy of each step and calculation, trying to find what I assumed was a mistake. What had I missed? Where had I gotten off track? After a lot more thinking and many more walks in the solitude of the hot and humid North Carolina trails, I finally realized what I was missing. It turns out I was missing a lot!

In my first try, I understood how the branches of the wave function of the universe localize on the various landscape vacua.

But I had missed a crucial piece of the puzzle: the separation or decoupling of the different entangled branches of the wave function of the universe from one another as they produce infant universes. Including entanglement among the branches of the wave function in my equations was not the final chapter in calculating the probability of our origin. To complete the project, I had to identify a way to decouple the entangled branches in the wave function of the universe as they were about to inflate and create their own universes.

Unlike entangled quantum particles, big classical universes cannot add up, interfere, or become entangled under the same sky. Entanglement, as we will see in more detail in the next chapter, is a purely quantum effect that doesn't exist among classical objects. Thus, it needs to be wiped out before our quantum universe, and the others entangled with us, can grow and make the transition from microscopic quantum wave to macroscopic classical universe. In physics, this process of decoupling, the mechanism that destroys entanglement, is known as decoherence.

Overlooking decoherence had taken me full circle back to the mystery of our origin—the solutions I found incorrectly indicated that the most likely universe is one that starts from the lowest Big Bang energy.

Decoherence would turn out to be the key to our final result—that, in fact, the most probable universes produced out of the quantum landscape were universes that started out at very high energies, just like our own universe had! Our result demonstrated, first, that in contrast to Penrose's estimate, our origin is a very likely one—there is absolutely nothing special or exclusive about our beginning; and second, that the story of the origin of the universe could now be calculated and derived using the laws of nature, and not just postulated.

It was a thrilling moment—but I wouldn't be able to dwell on it for long. I was about to discover that a quantum selection process was in fact at work, one that gave those infant universes unequal chances of existence—and that helped explain how our universe turned out the way it did.

9

The Origin of Our Universe

I N   H I S   M O N U M E N T A L   theory about the many worlds of quantum mechanics, Hugh Everett argued that all the branches of the wave function have equal chances of existence as universes. However, my derivation showed that not all universes produced from the branches of the wave function trapped in the landscape of string theory have equal chances to come into existence. In fact, I found that for some, the chances of existence are nearly zero.

There are two factors that play an essential part in the selection criteria and the survival chances of the universe. The first one is the energy these quantum wave packets "borrow" from the landscape vacua, the energy that initiates an inflationary growth in the infant universes. But there is more to the story.

The second factor, which plays an equally important role in determining whether the inflationary growth happens and if a universe will come into existence from its respective wave packet, is the amount of quantum fluctuations present inside each wave-universe.

Quantum fluctuations are present in the wave function and on the landscape vacua left over after compactification (the process used to get the eleven-dimensional string theory down to four dimensions). Because we are treating these proto-universes as quantum objects, all of them will unavoidably contain quantum fluctuations. I had not accounted for these fluctuations in my previous calculation, but I realized that they must have played a crucial role in the creation of our universe because they are equivalent to matter particles.

The emerging picture of these quantum universes had them contain two ingredients: the energy of the landscape vacua where they were localized, and matter particles in the form of quantum fluctuations. And due to another unavoidable aspect of quantum theory, branches in the wave function of the universe—just like any group of quantum particles—will engage in a type of quantum cross talk, where they appear to communicate instantaneously with each other. We call this interaction quantum entanglement.

Today, quantum entanglement plays a pivotal role in studies of neural networks and the mind and in the development of quantum computing, quantum information, and artificial intelligence. But in theoretical physics, entanglement proved to be central to the understanding of our origin from the quantum multiverse. As we will see, entanglement is also what finally, and surprisingly, enabled the scientific testing of our theory about those very origins.

Historically, quantum entanglement bothered Einstein more than any other aspect of quantum theory, as he firmly believed in only one objective reality and wanted to know what all this mess of possible entangled universes meant for the real world. He maintained to the end that a major piece was missing from the quantum puzzle, especially in light of what he derided as the "spooky action at a distance," an instant speed of communication resulting from entanglement. To make his case, in debating with Bohr, he did what he could do better than anyone—he came up with a series of thought experiments that produced paradoxes for quantum theory.*

These debates were an amazing intellectual feat between two giants of twentieth-century physics. Both respected truth and each other's opinion. In a letter to Bohr in 1920, Einstein wrote, "Not often in life has a human being caused me such joy by his mere presence as you did." Each time Bohr was pushed into a corner by Einstein's thought experiments, he had to think just that bit harder to figure out the answer and win the argument. Quantum theory advanced further an inch at a time.

The issue of "spooky action at a distance" and Einstein's paradoxes were ultimately settled by the realization that entanglement relied heavily on the dual nature of quantum particles as waves. Being strictly quantum in nature, these entangled particles could not transfer classical information. Since the phenomenon of entanglement was confined exclusively to a quantum world and did not exist in the classical world, the speed-of-light limit for exchanging information in our classical world was safely preserved.

So how does entanglement work in a quantum world? An easy way to visualize this is to think of a familiar particle: an electron. Electrons have a property called spin that is a purely quantum-mechanical effect; spin does not have a counterpart in the classical world we are familiar with. One way to picture the spin of an electron is as a rotation around its axis. The amount of spin on a particle is its quantum identity; like a birthmark, it

---

* The most famous of those directly aimed at entanglement was the EPR paradox, named after Einstein, Boris Podolsky, and Nathan Rosen. Rosen was part of the theoretical physics group at UNC Chapel Hill when the EPR paradox was published in 1937.

cannot change. Different quantum particles have different spins. Electrons have a choice of a spin up or a spin down.

Now imagine an atom that has two electrons and, thanks to some symmetry, is in a state of zero total spin—that is, if one electron is spin up, then the other must instantly flip to spin down; if one spins clockwise, the other one must spin counterclockwise. The spin of the two always adds up to zero. Crucially, this symmetry exists whether the electrons are in the same orbit around the same atom or are spread across the universe.

How do the two electrons know about each other's spin orientation and organize themselves to opposite spin directions at all times, no matter their location? There is only one way: Somehow the pair of electrons are instantly "communicating" this information about their spins to each other so that if one electron flips spin up, then the other automatically knows about it and flips spin down instantly. This interaction among quantum particles is quantum entanglement, and in our example, the two electrons are entangled. Like a strong marriage or a pair of twins, once two quantum particles are entangled, they remain entangled for life; if we were to separate our pair of electrons by vast distances—say, by placing one electron on the surface of the sun and the other on Earth—they would continue to be entangled and maintain instant communications about their spins.

If this feature seems odd, that's because it is. Instant communication of information (from one electron to the other, in this example) across vast distances implies infinite speed for the information to travel, whereas we know, thanks to Einstein, that nothing in nature can travel faster than the speed of light. Indeed, this is why Einstein objected to this feature of quantum theory, disparagingly calling it "spooky action at a distance."

Despite its appearance, however, quantum entanglement does not in fact violate Einstein's speed-of-light limit in nature. In our scenario, no classical information is traveling with infinite speed. To see how this can be possible, recall the wave-particle duality of quantum mechanics. Because of it, quantum particles (whether electrons or quantum infant universes) are not simply point-like objects that exist in one location or another; they are also waves that spread all the way to infinity. Thus, two different far-flung quantum particles need not travel to meet each other (or transmit information across the distance between them); they are in contact with each other over vast distances at all times. They are quantum entangled.

Recalling that quantum entanglement operates purely in the quantum realm, I realized that, although I started my derivation of the origin of our universe at a time when we were just a quantum wave-universe, a branch of the wave function of the universe on the landscape of string theory, somehow this theory needed to wipe out entanglement with the other branches, to produce a classical universe like the one around us today, and do so in a coherent way.

In order to achieve this quantum-to-classical transition for our universe, I needed the effect of a second piece in my theory: quantum fluctuations, which have a major role in determining where potential universes get stuck on the landscape and where and how they survive.

These neighborhood fluctuations were a reminder of why it was important to take them into account when exploring the microscopic processes of the wave function. The impact of quantum fluctuations on each wave packet and on the landscape vacua they occupied completely changed the odds of producing a universe

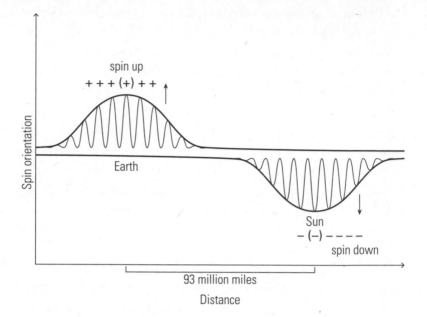

Figure 12. Entanglement of two electrons (one thick line, one double line). As waves, both electrons spread to infinity. If they were point particles, they would need to send light signals across the distance between them—in this case, the distance from the sun to the Earth, which would mean that each signal would take eight minutes to travel from one to the other. But as waves, they are constantly in contact, so there is zero distance they need to travel to communicate and therefore no delay in their exchange of information. Thus, when one goes spin up, the other one knows of it instantly and flips spin down; in the same manner, should one spin go to the right, the other will go to the left instantly. This happens without breaking the speed-of-light limit.

out of these wave packets. At the same time, adding an infinite number of fluctuations into an already complicated set of mathematical equations for how the wave function of the universe passes through a vast landscape seemed like an impossible task.

Ultimately, my curiosity won out. But I was aware this was too much work to handle alone. I knew the perfect person to help—he was not only an expert in quantum cosmology but also a delightful person with a great sense of humor. So one morning after I finished my classes, I called Rich Holman in the physics department at Carnegie Mellon University. I was counting on Rich to provide a second pair of eyes and to play devil's advocate for any blunder or mistake I might accidentally make. Rich was intrigued, and what started as a single project turned into many years of enjoyable and fruitful collaboration. At various points when we thought we were stuck, Rich lifted the gloom with one of his jokes and his cheerful attitude.

We began the daunting task of repeating my previous calculation of the wave function of the universe on the quantum landscape but this time taking into account an infinite number of fluctuations weakly coupled with our branch in addition to the nearly infinite number of landscape vacua. The collection of these fluctuations filled the landscape and "watched" the branches of the wave function as they spread through the landscape. (Think back to Hugh Everett's work on Schrödinger's cat-in-the-box thought experiment in which he concluded that as quantum objects, both the cat and the observer are able to constantly watch each other.)

But classical universes cannot be entangled quantum particles. Therefore, quantum universes that begin as entangled waves in a universal wave function of the universe, in a state of constant communication, have to separate from each other on the great universe-factory chain and acquire their own independent individuality in order to become classical universes. They break free from entanglement through a process called decoherence.

Decoherence destroys the kinship among the wave packets and makes them decouple from one another as they transition from quantum waves to inflating universes. It washes out entanglement among quantum particles and, with it, their quantum nature. Decoherence is how they transition from being quantum objects to becoming classical universes. Think of it as an irreversible border crossing between the two worlds: a quantum world, where the entropy is zero and the governing rules are different, and a classical world, where entropy grows and time moves irreversibly forward.

Microscopically, decoherence is triggered by the interaction of the wave function with the environment, or the "bath" of quantum fluctuations, in which it is immersed. The interaction with the bath pins them down to single values and locations, thereby forever destroying their inherent quantum uncertainty.

One way of understanding decoherence is to think of the process of separating pure gold from ore. Placing gold ore in a hot bath of borax causes many of the different minerals in the ore to melt. Each mineral reacts differently to the borax bath. As their melting temperatures are reached, these minerals separate from one another, and the gold, which experiences the least interaction with the borax, sinks to the bottom.

In our case, the ore is the wave function of the universe mixed with all the branches. The bath in which this wave function is immersed and with which it interacts is the collection of quantum fluctuations that are trying to stop the universes from growing. As a result of this entanglement with the fluctuations bath, branches of the wave function separate—that is, they decohere—from one another while they are also taking energies from the landscape vacua in order to undergo their own Big Bang inflation.

Decoherence occurs during the brief instant when the branches of the wave function of the universe located on the energy sites of the landscape are about to go through their individual cosmic inflation processes to grow and produce individual universes.

Once decoherence is completed, each growing universe acquires its own individual identity and is free of quantum uncertainty; it becomes decoupled from the other growing universes, producing its own space-time independently of similar processes in other developing universes.

I didn't realize then the impact that accounting for the bath of fluctuations and decoherence would have on deriving the probability of our origins. It turned out that this would be the key to solving the mystery of our universe's unlikely origin.

But decoherence and entanglement are two sides of the same coin. Wave packets of the universe that settle in high-energy sites along the landscape can survive the squeeze of fluctuations' attractive gravity with only a few "dents"—minor changes to their original shape—and continue through the phase of rapid cosmic inflation to produce a macroscopic universe like ours. Since the separation—decoherence—of the branches of the wave function of the universe (which become universes) is triggered by their entanglement with the environmental bath of fluctuations, I further wondered if we could calculate and find any traces of this early entanglement—the cross talk among all the branches that produced universes—imprinted on our sky today. If I could rewind the creation story all the way back to its quantum land-scape roots, when our wave-universe was entangled with others, then the moment of decoherence is where we could follow the evolution of the universe to the other side, when our universe was just a quantum wave packet, beyond the reaches of our classical world.

As we worked back through the calculation of the wave function of the universe on the quantum landscape, now factoring in an infinite number of fluctuations as well as a nearly infinite number of landscape vacua, Rich and I discovered what we had been hoping to find all along. Interaction with the bath of quantum fluctuations triggered the branches of the wave function to decohere from one another without changing their properties in the process of measurement. Our next step was to calculate the net effect these fluctuations had on the evolution of each of these individual branches when they became classical universes. Considering the complexity of such a system, we had no way of knowing or anticipating what the solutions would be. But after half a year of intense work, what we found proved worth the effort.

When the branches of the wave function evolved to become individual universes, we discovered, they did so by taking the energy from different energy vacua where they had settled on the landscape. Then these branches went through the process of decoherence to disentangle from each other, acquire their own individual identity, inflate and become growing classical universes.

What we discovered was that infant universes that started at very high energies were the most likely universes to be produced out of the quantum landscape, just like we knew from astrophysical observations and the theory of cosmic inflation to be the case in our own universe! For the first time, our results demonstrated that, although our universe was not eternal, it had a moment of creation 13.8 billion years ago, and there was absolutely nothing special about it, and, just as thrillingly, that the origin of the universe could be calculated and derived instead of supposed. Penrose's conclusion on the improbability of our origin turned out to be incorrect, once we figured out how to peer behind the curtain of space-time and glimpse the larger canvas of the cosmos, the multiverse from which our universe had been born. The mystery of the improbability of our universe's existence disintegrated once you looked at the bigger picture and the dynamical processes within it, the vastness of the quantum landscape multiverse, where universes were constantly being created as long as they had sufficient energy.

Furthermore, in contrast with Everett's picture of the many worlds of quantum mechanics, we found that not all the branches of the wave function of the universe have an equal chance to produce a universe; the probability of each branch becoming a universe depended on a dynamical selection rule: the energy of the landscape on which the branch had localized and on the amount of quantum fluctuations captured within it. In the absence of a selection criterion, Everett had assumed that each branch of the wave function had an equal chance of bringing a universe into existence. The concept of decoherence was formally discovered more than two decades after Everett wrote his dissertation. Since he had no way of assigning weights to the branches of the wave function of the universe, he based his theory on what is known as "the principle of ignorance"—if we cannot estimate the chance each universe has to exist, then we should assume that all universes, both high-energy and low-energy, have equal chances of existence. The emerging selection mechanism for the infant universes that start from landscape energies that Rich and I derived showed that this is not the case. For some universes that start at low energies, the chances of existence are nearly zero. And for many others, a whole multiverse of high-energy ones like our universe, the chances of success are very high. Only the fittest infant universes survived. If you recall, the probability was simply the square of the solutions found; therefore, different solutions for the branches of the wave function correspond to different probabilities for producing a universe.

Physically, fluctuations behave like matter particles, meaning they would take that tiny quantum universe and try to crunch it under their weight into a black hole. (In physics, we say that matter has a positive heat capacity, meaning it possesses attractive gravity.) In contrast, energy would explode that initial universe by driving it into a cosmic inflationary phase. (In physics, we say that the gravitational field of energy, unlike matter, is a negative heat-capacity system, meaning it has repulsive gravity.) Each branch of the wave function of the universe contains the energy it has taken from the landscape valley where it is confined and the energy it has taken from the particles in the form of fluctuations. And this is how the infant wave-universe ends up in a wild tug-of-war: a fight for dominance between the energy the branch took from the landscape vacua on which it settled that is trying to drive it through an accelerated expansion and the pull from the fluctuations behaving like matter inside them trying to stop growth and crunch them into something roughly resembling a black hole.

How does this process work? When I talked about Big Bang inflation and dark energy, I explained that dark energy takes a tiny quantum universe and tends to inflate it very quickly. But recall that matter has attractive gravity—it takes that initial quantum universe and crunches it into a point, compressing its contents, as happens inside black holes. We know from Einstein's equations ("matter and energy tell space how to curve") that this is what happens to a tiny universe filled with matter or energy. Each of these branches of quantum universes that are still wave packets, localized somewhere on the landscape's valleys, contain both matter (which, through quantum fluctuations, is trying to stop expansion of the quantum universe) and energy (which, taken from the landscape, is trying to drive the quantum universe into inflated growth). Whichever one is stronger in their tug-of-war determines if a universe will be born out of that particular energy site on the landscape.

This means that, contrary to previous estimates, which gave the smallest chance of existence to our universe, universes that start inflating at high energies—as our universe did—have the *highest* chance of coming into existence and growing into macroscopic universes. Wave packets that settle at low-energy sites in the landscape cannot grow; they remain squeezed to their quantum size and can never produce an observable large classical universe. They become what we call terminal universes, remaining microscopic quantum wave packets for eternity. We could in fact calculate the survival chances a wave-universe had in producing a big classical universe depending on which vacua it had settled on.

A selection mechanism for the creation of universes from the landscape quantum multiverse was at work—the competition inside each branch between the behavior of matter (contained in the fluctuations) and gravity (as determined by the landscape energy that drives the universe through Big Bang inflation). Through this tug-of-war, nature displayed its own version of Darwin's natural selection: it gave infant universes unequal chances of survival and even wiped out the terminal universes from existing in the classical world.

As in the case of a Schrödinger-like equation in quantum mechanics, the Wheeler-DeWitt equation for the wave function of the universe on the string-theory landscape gives not one but many solutions of infant wave-universes, each of their branches localized in different vacua of the landscape (some on high and some on low energies). Therefore, even though some of those solutions—the terminal universes—are removed in our theory, the number of surviving universes that start out from high-energy landscape vacua and grow is still very large; they produce an entire quantum multiverse.

By including gravity and fluctuations in our equations, we transformed the string-theory landscape into a "fitness landscape." Only the fittest universes, the ones that settled in high-energy vacua in the landscape and inflated at high energies, survived, grew, and produced macroscopic universes.

This is the key to what we had discovered: The odds of our existence had changed! This time, solutions for the most probable universes produced out of the quantum landscape started out at very high energies, as our own universe had! Our derivation of the chance for our existence demonstrated that there is nothing special or fine-tuned about the origin of our universe; our universe's chance of existence is high simply due to evolutionary selection, determined by the quantum dynamics of gravity and matter.

The week after we finished our calculations was an emotional roller coaster. The day after we finished, Rich and I were stunned at having managed to do all the calculations. The next day, we might have been the happiest people in the country, and we couldn't stop smiling. And on the third day, we were crushed with doubt, reminding ourselves that the enticing answer we thought we had found was such a huge claim, it was probably wrong for reasons we hadn't yet identified.

If I'd learned any lesson from reading about the history of scientific ideas, it was humbleness.

I recalled one of the stories my dad told me about Enrico Fermi, the Nobel laureate physicist and architect of the atomic bomb whose old office had been on the same hallway as mine when I was a postdoctoral fellow at Scuola Normale Superiore in Pisa. As my dad told it, despite the excitement of being close to finishing the atom bomb, that last evening Fermi asked his team to stop work on their nuclear calculations and take a break.

The moral of the story, my dad explained, was that you should exercise caution and restrain your excitement. Don't rush important things. It is easy to make a mistake when you've almost reached the end of a scientific marathon and are tired. As hard as it is, walk away from it for a while. Then, once you have detached from it, return with fresh eyes, double-check, triple-check, and check again. A week after Rich and I finished our calculations, that is what we did. After that, we put the project away for about a month, then rechecked the calculations before sending it to a journal and posting it on the online physics archives.

When we reached the end of our calculations, Rich and I felt simultaneously elated and nervous. We had found a way to mathematically derive the answer to the origin of our universe and of the world that lies beyond its space and time boundaries. We believed that we were correct. But would the rest of our universe agree?

At the same time that Rich and I were celebrating our discovery, we were also giving serious consideration to the other, alternative answers to the landscape crisis that were emerging thanks to work being done by our colleagues in the physics community.

One of the alternative approaches to avoiding the "landscape crisis" was to add an intellectual asterisk to the string-theory landscape. This approach rationalized and limited the choices offered by the landscape of string-theory discovery by using something called the anthropic principle.

The anthropic principle argued that observers actually picked which universes were viable to live in to allow for their habitation. Anthropic-principle reasoning is very much like a twenty-first-century restatement of Bohr's collapse of the wave function, or (reaching further back) a rephrasing of Descartes's famous aphorism "I think, therefore I am." These efforts to solve the landscape crisis were essentially advocating, "I think, therefore the universe exists," with its observer of the universe, like us, always capable of witnessing its existence. But rather than selecting the one "true" quantum particle, we were selecting the one "true" universe and discarding the rest of this string-landscape vastness as irrelevant.

The anthropic principle is more philosophical than scientific and can best be expressed by this logical tautology: This universe exists and is the true universe because we, as sentient beings, can see and witness its existence. We came to be here through a long and complex process that involved galaxies and star factories. And stars would not exist had it not been for the fine-tuning of the constants of nature and the fine-tuning of cosmic inflation to produce our universe today. Thus, for life to arise, the universe had to begin in a very special state with all its ingredients and forces fine-tuned. Therefore, even if our universe had a ridiculously small chance to exist, as Penrose estimated, it is still the only one with the right conditions for structure and, ultimately, life to arise, so it is no surprise that we find ourselves in it.

Fine-tuning has a very specific connotation in physics—it means that the constants of nature, like the mass or charge of an electron, which determines the strength of forces of a universe, have to be set to an exact numerical value, the value they take in our universe. Even a small change in only a few of the equations and values for these laws of nature would result in the creation of a radically different universe—or no universe at all. Thus, using the existence of life and saying that life could only exist if the universe was fine-tuned as the criteria for selecting our present universe seemed to anthropic-principle supporters to resolve the "landscape crisis." It offered a picture for how one "real" universe could emerge out of the vast number of possibilities.

Additionally, according to the anthropic principle, the *only* virtue of having a vast landscape, or any type of multiverse, for that matter, was to increase the chances that a fine-tuned universe like ours would be found. It presumed that only very few universes—preferably one—like ours are available. It further conjectured that life could only exist in a universe exactly like ours, a universe with exactly the same special constants of nature, Big Bang inflation, and dark energy. I was skeptical of the validity of these assumptions. Anthropic selection started off by assuming the very answer we wanted to hear and then tried to rationalize its choice.

At the time, the anthropic principle was embraced by many of the titans of theoretical physics, many of whom I greatly respected. But to me, it seemed like giving up on solving the problem. The more I thought about it, the more I believed that the anthropic principle actually unleashed a new set of questions, starting with the very basic "Why did this happen and not that?" The world-renowned British physicist Paul Davies in his book *The Mind of God* refers to this problem as the "God-of-the-gaps,"* which he defines as the special divine-intervention scenario that we humans employ in order to address a gap in our understanding or explanations. The modern origins of this "God's-eye view" date to early-seventeenth-century France where, as I have noted, the French philosopher, mathematician, and scientist René Descartes expounded on his belief that human beings have a higher moral status than every other organism and are the preeminent species on Earth. Thus, he argued, reality can be assessed only through a human perspective. In positing his premise of *cogito ergo sum* ("I think, therefore I am"), Descartes declared his strong advocacy for the power of human reasoning as the way to understand the workings of nature and the cosmos. In doing so, Descartes also pioneered what became known as the scientific method. In many ways, this legacy of Cartesian dualism, the belief that "I think, therefore the universe exists," unfortunately seemed to underlie the reasoning behind the anthropic principle in cosmology.

But everywhere I looked, the ranks of supporters for the anthropic principle were growing; in fact, the Nobel laureate

---

* The term *God-of-the-gaps* was first used by the Oxford mathematician C. A. Coulson in his 1955 book *Science and Christian Belief.*

Steven Weinberg had used anthropic arguments to make a case for the existence of dark energy before it was observationally discovered, a prediction that only increased enthusiasm about the anthropic principle among its supporters.

The anthropic principle certainly seemed to solve the issue of the lack of testability for a theory about the universe by placing us center stage as its observers. This meant that the only viable universes were the ones like ours; if another universe did not have the right conditions for stars, galaxies, and, ultimately, life to arise and bear witness to its existence, then it might as well not exist. The problem I had with this approach was that indirectly, scientists had already decided on the answer they wanted— namely, the only "good" universe was ours.

As the multiverse research gained momentum, so did the elaboration of the anthropic principle. Its proponents justified the multiverse as the perfect setting for increasing the chances to find a universe like ours. I was not convinced that these arguments were a scientifically derived answer to our origins enigma. Perhaps my reluctance had something to do with the fact that this new spin on the multiverse—a spin meant to reduce the possibility of many universes to a single one like ours sounded familiar. Among other things, it reminded me of a sad event from my dad's life, a moment when he had his own universe's possibilities dramatically curtailed.

In my dad's day, the highest recognition an Albanian student could receive was a scholarship to study in the Soviet Union, and competition for scholarships was fierce. Each year, the best student at every high school was awarded a gold medal at graduation. A gold medal guaranteed a scholarship to study in Moscow.

My dad, despite his family's many difficulties, excelled academically, often completing two grades in one year. He far exceeded the standards required for the gold medal. His teachers, especially his math teacher, assured him that the honor was a foregone conclusion. My dad even took it upon himself to learn Russian to be ready for his scholarship.

Graduation day arrived. Just as the principal was about to hand my dad the medal, the school's Communist Party secretary snatched the medal away and gave it to another student, who was as shocked to receive it as my dad was to lose it.

Later, my dad would joke about this incident and say that it was a blessing in disguise for two reasons. First, having learned Russian, he now had access to all the literature of the West, since no one in the Albanian government had thought to ban Russian translations of works of science, history, art, or music. Second, many Albanian gold-medal students returned from their studies with Russian wives, but after Albania severed relations with the Soviet Union in 1967, all the Russian wives and their children were rounded up, put on a plane, and deported to Russia. Most of the Albanian fathers never heard from or saw their wives or children again until the 1990s. The very few men who refused to break up their families were sentenced, along with their wives and children, to forced-labor camps. My dad said he could not have survived losing his children like that.

But my father also could not stop talking about the medal. In his later years, he even wondered if we could request that the Ministry of Education dig through its archives and find the documentation proving he had won the medal. A misappropriated prize from a long-ago high school graduation. The odds of it being found seemed low, to put it mildly.

For a long time, I could not understand for the life of me how someone like my dad, who more than once gave his only jacket to a stranger on the street when he thought that person was colder than he was, could be so obsessed with a medal. This was a man who seemed to care nothing for possessions, who, to my mom's horror, and without hesitation, gave his life savings to a friend and later gave his entire monthly paycheck to a student to pay for his wedding because the young man's own family was so poor. He was a man who wouldn't even know if he was wearing different-colored socks if my mom didn't check his clothing daily, a man to whom material possessions and false praise meant nothing. A man with no vanity and no bitterness—how could a gold medal from his teenage years mean so much?

When I asked my father why the medal was so important to him, he candidly explained that he had sacrificed a great deal, living with few clothes and not enough food, in order to attend school, but he had looked up to his teachers, and learning was what kept him going. He saw his adult experiences of being mistreated and betrayed and backstabbed by friends and colleagues as part of human nature, one group of adults hurting another one out of envy or professional jealousy. The Communist system then did the rest of the damage. But as a child, he had the greatest respect for his teachers. He maintained that teachers had the noblest role in caring for the well-being of children, and only the cruelest of societies would force teachers to hurt an innocent child.

After my father lost the medal, his math teacher stood by him and consoled him by telling him that there was another prize that everyone wanted. It was a prize that no one could take away and no money in the world could buy: the power of his own mind. But those words of encouragement hadn't quelled his deep disappointment.

Despite my father's explanations, it was only when I became a parent myself that I began to understand the impact of this experience on him. I could not imagine such an experience happening to my daughter. To work so hard only to have her just rewards snatched away by those she trusted the most—it would be devastating.

In my own life, I had been given the chances that had been denied to my father. I needed to make the most of those opportunities, not simply follow the easier path, even if that meant challenging some deeply held beliefs of my own.

If I was going to make an affirmative case for the multiverse, I needed to scrutinize anthropic reasoning. Working with my collaborator and friend Fred Adams, the distinguished astrophysicist at the University of Michigan in Ann Arbor, I decided to investigate how anthropic selection and the related fine-tuning of the universe might work in practice. As an astrophysicist, Fred knows a lot more than I do about the astrophysical aspects of stars and galaxies, how these structures formed in the universe, and how stars became factories that produced elements heavier than hydrogen. His expertise would allow us to find out if structures would form in universes that had very different conditions than ours, such as different strengths of gravity or electromagnetic forces.

To understand the basis for anthropically selecting a universe, we posed a series of questions: What does a habitable universe look like anyway? Is it a universe where laws are universal and where the fundamental constants of nature—such as Newton's gravitational constant, electron charge, and proton mass—are fixed and fine-tuned to maintain exactly the same values that we observe in our universe and that we know have allowed life to arise here? Is a habitable universe one that contains dark energy—which, Weinberg found, might be necessary to produce habitation?

We quickly realized that if a universe contained dark energy, it would not be habitable in the far future. It would be difficult to justify the existence of such a universe using anthropic arguments. A universe that contains dark energy ends up empty, cold, lightless, barren of life, and incapable of creating new structures because it is not allowed to change its entropy state, resulting in a cosmic heat death of all its structures and observers. It may have a relatively brief time available for habitation, but after that, it spends eternity in the empty state of heat death. It is precisely the wrong scenario to allow life to hang around long enough to witness the creation of a flourishing universe.

Later on, Fred and I and two other colleagues, the pioneering physicist Stephon Alexander at Brown University and Evan Grohs, then at the University of Michigan, expanded our investigation into the merits of the anthropic arguments. We inquired whether the fine-tuned values of the constants of nature observed in our universe provided the only possible conditions for life. For life to arise, we need a handful of requirements, including a certain amount of complexity of at least 10^{15} particles in the universe and long-lived stars that act as factories to produce heavy elements from lighter elements under the gravitational pressure inside the star's core.

The result of our investigation surprised even us: Habitable universes could exist even if we made the strength of gravity a lot weaker or a lot stronger and even if we changed other constants of nature (like the constant that controls the strength of electromagnetism) by many millions of times from their known values. We concluded that the constants of nature in our universe are not specially selected to allow for habitation. Even worse for the anthropic argument, we found that our universe seemed only borderline habitable based on the anthropic selection rules. There were many other possible universes with very different constants of nature from ours that would be more likely to allow life to arise.

These findings made me even more convinced that the application of the anthropic principle was like throwing in the towel on science. However, support for an anthropic selection of our universe from the landscape was continuing to increase. In a new twist, the string-theory landscape was being retooled as an elaborate justification for anthropic selections; the landscape's vastness increased the chances of finding an anthropically fit (that is, habitable) universe, while the rest of the possibilities became redundant.

I thought this way of thinking was dangerous; I believed that the existence of our universe could not and should not depend on who observed it. How could we simply select our origin on the basis of our own habitation? It increased my motivation to demonstrate that the answer to our universe's origins did not require anthropic reasoning but could be derived through physics equations and the laws of nature.

In fact, I came to believe that the answer to the question of how we can scientifically glimpse our origins was staring us right in the face—it was in the skies above.

After Rich's and my initial excitement over reaching a solution, reality sank in. We had a promising theory that, for the first time, derived the answer to the long-standing enigma of our universe's small chance of existence. But it was a theory of the multiverse, a topic that was still beyond the pale to many physicists. To convince the scientific community—and ourselves—that we were on the right track, we had to find ways to test our theory. We had to demonstrate that the multiverse could in fact be subjected to scientific scrutiny and prevail.

The scientific community had long held a deep conviction that testing the multiverse was impossible due to the speed-of-light limit, which permits us to observe only objects within the horizon of our universe. But Rich and I were not discouraged. Our theory that our universe originated from the quantum landscape multiverse was based on mathematical derivations using the Wheeler-DeWitt equation of the formalism of quantum cosmology—the only valid theory of quantum gravity at the time when our universe was a quantum particle seething with high energies. Recall that we used the Schrödinger equation when calculating the paths of, say, electrons moving in real space-time under the influence of some potential energy; the Wheeler-DeWitt equation is the Schrödinger equation's equivalent in cases when quantum particles, such as the electrons in our example, are branches of the wave functions, meaning when they are infant wave-universes moving on an abstract space of energies instead of on physical space-time. The Wheeler-DeWitt equation's solutions, just like in quantum mechanics, are proportional to the probabilities that the infant universe takes a particular path on that abstract space of energies.

Of course we were aware that so far our results were highly theoretical and that being taken seriously by our community required us to find ways to test it. But ours was a theory of the origin of our universe from a quantum multiverse, which brought us face to face with the most difficult issue: How could a theory of the multiverse be tested?

Scientific investigations of problems like the creation of the universe, which we can neither observe nor reproduce and test in a lab, are similar to detective work in that they rely on intuition as well as evidence. Like a detective, as pieces of the puzzle start falling into place, researchers can intuitively sense the answer is close. This was the feeling I had as Rich and I tried to figure out how we could test our theory about the multiverse. Rationally, it seemed like a long shot, but intuitively, it seemed achievable.

Finally, a potential solution hit me. I realized that the key to testing and validating this theory was hidden in quantum entanglement—because decoherence and entanglement were two sides of the same coin! I could rewind the creation story all the way back to its quantum-landscape roots, when our wave-universe was entangled with others.

I already knew that the separation—the decoherence—of the branches of the wave function of the universe (which then become individual universes) was triggered by their entanglement with the environmental bath of fluctuations. Now I wondered if we could calculate and find any traces of this early entanglement imprinted on our sky today. Could we look for vestiges of the landscape era—evidence of cross talk, quantum communication between all the branches of the wave function of the universe that had produced individual universes?

This might sound like a contradiction. How could our universe possibly still be entangled with all the other universes all this time after the Big Bang? Our universe must have separated from them in its quantum infancy. But as I wrestled with these issues, I realized that it was possible to have a universe that had long since decohered but that also retained its infantile "dents"—minor changes in shape caused by the interaction with other surviving universes that had been entangled with ours during the earliest moments—as identifiable birthmarks. The scars of its initial entanglement should still be observable in our universe today since it is simply a blown-up version of its infant self.

The key was in the timing. Our wave-universe was decohering around the same time as the next stage, the particle universe, was going through its own cosmic inflation and coming into existence. Everything we observe in our sky today was seeded from the primordial fluctuations produced in those first moments, which take place at the smallest of units of measurable time, far less than a second. In principle, during those moments, as entanglement was being wiped out, its signatures could have remained stamped on the inflaton and its fluctuations. There was a chance that the sort of scars that I was envisioning had formed during this brief period. And if they had, they should be visible in the skies.

Understanding how scars formed from entanglement is less complicated than you might imagine. I started by trying to create a mental picture of the entanglement's scarring of our sky. I visualized all the surviving universes from the branches of the wave function of the universe, including ours, as a bunch of particles spread around the quantum multiverse. Because they all contain mass and energy, they interact with (pull on) one another gravitationally, just as Newton's apple had its path of motion curved by interacting with the Earth's mass, thus guiding it to the ground. However, the apple was also being pulled on by the moon, the sun, all the other planets in our solar system, and all the stars in the universe. The Earth's mass has the strongest force, but that does not mean these other forces do not exist. In analogy, the net effect that entanglement left on our sky is captured by the combined pulling on our universe by other infant universes. Similar to the weak pulling from stars on the famous apple, at present, the signs of entanglement in our universe are incredibly small relative to the signs from cosmic inflation. But they are still there!

I will admit it . . . I was excited by the mere thought that I potentially had a way to glimpse beyond our horizon and before the Big Bang! Through my proposal of calculating and tracking entanglement in our sky, I may very well have pinned down, for the very first time, a way of testing the multiverse. What thrilled me most about this idea was its potential for making possible what for centuries we thought was impossible—an observational window to glimpse in space and in time beyond our universe into the multiverse. Our expanding universe provides the best cosmic laboratory for hunting down information about its infancy because everything we observe at large scales in our universe today was also present at its beginning. The basic elements of our universe do not vanish over time; they simply rescale their size with the expansion of the universe.

And here is why I thought of using quantum entanglement as the litmus test for our theory: Quantum theory contains a near-sacred principle known as "unitarity," which states that no information about a system can ever be lost. Unitarity is a law of information conservation. It means that signs of the earlier quantum entanglement of our universe with the other surviving universes must still exist today. Thus, despite decoherence, entanglement can never be wiped from our universe's memory; it is stored in its original DNA. Moreover, these signs have been encoded in our sky since its infancy, since the time after the universe started as a wave on the landscape. Traces of this earlier entanglement would simply stretch out with the expansion of the universe as the universe became a much larger version of its infant self.

I was concerned that these signatures, which have been stretched by inflation and the expansion of the universe, would be quite weak. But on the basis of unitarity, I believed that however weak they were, they were preserved somewhere in our sky in the form of local violations or deviations from uniformity and homogeneity predicted by cosmic inflation.

After discussing this possibility, Rich and I decided to calculate the effect of quantum entanglement on our universe to find out if any traces were left behind, then fast-forward them from infancy to the present and derive predictions for what kind of scars we should be looking for in our sky. If we could identify where we needed to look for them, we could test them by comparing them with actual observations. And for the first time, we could demonstrate that the multiverse could be tested.

Rich and I started on this investigation with help from a physicist in Tokyo, Tomo Takahashi. I first got to know Tomo at UNC Chapel Hill in 2004 when we overlapped by one year. He was a postdoc about to take a faculty position in Japan, and I had just arrived at UNC. We enjoyed interacting, and I saw the high standards Tomo maintained for his work and his incredible attention to detail. I knew he was familiar with the computer simulation program that we needed in order to compare the predictions based on our theory with actual data about matter and radiation signatures in the universe. In 2005, I called Tomo, and he agreed to collaborate with us.

Rich, Tomo, and I decided that the best place to begin our search was in the CMB—cosmic microwave background, the afterglow from the Big Bang. CMB is the oldest light in the universe, a universal radiation "ether" permeating the entire cosmos throughout its history. As such, it contains a sort of exclusive record of the first millisecond in the life of the universe. And this silent witness of creation is still all around us today, making it an invaluable cosmic lab.

The energy of the CMB photons in our present universe is quite low; their frequencies peak around the microwave range (160 gigahertz), much like the photons in your kitchen microwave when you warm your food—hence the name cosmic microwave. CMB photons in the present epoch "heat" our sky to temperatures of about 2.7 K, or -271°C. (Or, if you prefer, -455°F.) But although the CMB is extremely cold, it is not cold enough to escape observation. Three major international scientific experiments—the COBE, WMAP, and Planck satellites, dating from the 1990s to the present—have measured the

CMB and its much weaker fluctuations to exquisite precision. We even encounter CMB photons here on Earth. Indeed, seeing and hearing CMB used to be an everyday experience in the era of old TV sets: when changing channels, the viewer would experience the CMB signal in the form of static—the blurry, buzzing gray and white specks that appeared on the TV screen.

But if our universe started purely from energy, what can we see in the CMB photons that gives us a nascent image of the universe? Here, quantum theory, specifically Heisenberg's uncertainty principle, provides the answer. According to the uncertainty principle, quantum uncertainty, displayed as fluctuations in the initial energy of inflation, is unavoidable. When the universe stops inflating, it is suddenly filled with waves of quantum fluctuations of the inflaton energy. The whole range of fluctuations, some with mass and some without, are known as density perturbations. The shorter waves in this spectrum, those that fit inside the universe, become photons or particles, depending on their mass (reflecting the phenomenon of wave-particle duality).

The tiny tremors in the fabric of the universe that induce weak ripples or vibrations in the gravitational field, what are known as primordial gravitational waves, hold information on what particular model of inflation took place. They are incredibly small, at one part in about ten billion of the strength of the CMB spectrum, and therefore are much harder to observe. But they are preserved in the CMB.*

-----

* Primordial gravitational waves are not the same as those produced by black hole mergers recently observed by the LIGO experiment. While similar in their nature, they are different in their origins. The former were produced during the primordial era of cosmic inflation and are extremely weak; the latter are produced in the current universe and are relatively a lot stronger.

Cautiously optimistic, Rich, Tomo, and I set to work to predict the remaining scars from this early quantum entanglement, what we jokingly described as a search for "avatars" of the quantum landscape multiverse in our own universe—traces of other universes that we would find in our sky as anomalous signatures. This would enable us to bypass the boundaries nature imposes on us by the speed-of-light limit and peer into the multiverse. In our curiosity, we were like three kids with a box of chocolates that we couldn't wait to open.

Using quantum cosmology, we computed the strength of entanglement of all the surviving universes with our own, from its quantum landscape multiverse time. We added this contribution to the primordial fluctuations of cosmic inflation (which gave rise to both the CMB and all forms of matter) and fast-forwarded its location projections to the present-day universe. This enabled us to derive and forecast where and how, in our theory, the CMB and matter distribution in our present sky are tweaked and modified by earlier contributions from entanglement.

Unlike the uniform, homogeneous distribution of CMB and matter sourced by cosmic inflation, entanglement's contribution to our sky varied with distance—meaning it was not uniformly distributed across the cosmos. Therefore, although contributions from cosmic inflation and entanglement are blended in the same sky, they can be separately identified. In fact, when they were observed, the entanglement signatures on our sky became known as "anomalies" because they break, however slightly, our overall universe's uniformity and homogeneity. Therefore, they cannot be explained by the standard model of cosmology—cosmic inflation—for a single universe, the key prediction of which is uniformity and homogeneity.

In short, estimating how entanglement modified the early universe everywhere in space-time allowed us to predict their possible locations in our sky. These entanglement scars were our road map to the multiverse. These predictions had intrigued me most; they had the potential for making observational detection possible. However, none of us anticipated the surprises that were about to arrive from the observational front. Up to that time, we had assumed that the current observational power was not strong enough to detect the signatures that we predicted. In other words, we thought we would likely not know whether our theory was right or wrong in our own lifetimes.

It turned out that the observational imprints entanglement leaves on our sky are strong enough to be detected. They trigger very specific, mild deviations from the uniformity and homogeneity of cosmic inflation. Through the entanglement scars on our sky, we opened a window that allowed us to see and test a rich world beyond the horizon of our universe—the multiverse.

10

Fingerprints of Other Universes

RICH, TOMO, AND I finally worked up the courage to sub-mit our results for publication in 2005 in a paper we titled "Avatars of the Landscape." Then we held our breath, waiting to see the reactions among our colleagues. And for a while, it was radio silence. As we waited, lines from William Blake's poem "Auguries of Innocence" kept playing in my mind: "To see a World in a Grain of Sand / And a Heaven in a Wild Flower / Hold Infinity in the palm of your hand / And Eternity in an hour."

Our calculations had led to several anomaly predictions. First among them was the existence of a giant void. We believed that resting in the distant sky above the Southern Hemisphere there was a primordial giant void, a hole signifying an almost empty region where the stars and galaxies are mostly "scooped out." On CMB-temperature maps of the sky, over-dense regions full of matter appear as hot spots, while empty regions, or voids, show up as cold spots. Normally, a CMB-temperature map of the sky shows a uniform sprinkling of small hot and cold spots, based on the uniformity feature of cosmic inflation (even though the hot and cold spots are randomly scattered, when they are all compared to one another, their distribution ends up being balanced, making them uniform).

But the primordial cold spot we predicted was different—it was huge! It covered an area of ten degrees, or roughly one-tenth of our visible sky, which is at least ten times larger than any of the normal CMB hot and cold spots originating from cosmic inflation. This was a gargantuan structure, and we named it the Giant Void. We had a mathematical derivation predicting its existence that showed it to be a scar left over from our universe's birth. We also predicted it would be about ten billion light-years away.

It was a bold claim, and naturally, we were nervous. Such a giant void clearly broke the uniformity principle of cosmic inflation. Indeed, at that time, talking about a giant hole in our sky and positing its relation to the multiverse did sound ridiculous. It seemed to be the surest way for future observations to prove us wrong.

But we expected that those observations would take awhile to be collected. We thought our telescopes and other observational technology here on Earth and even the satellites in space would not be strong enough to see into the universe with enough detail to confirm (or contradict) our predictions. As confident as we were, we were almost certain that we would not know whether our theory was right or wrong in our own lifetimes. None of us anticipated the surprises awaiting us.

Half a year after we published our first paper, "Avatars of the Landscape," with a list of predictions for what and where the anomalous scars were in our sky, a team of radio astronomers at the University of Minnesota accidentally spotted exactly the sort of giant void that we had predicted at precisely the size and distance we had predicted. The Giant Void was observed again, two years later, in the CMB maps created by the WMAP (Wilkinson Microwave Anisotropy Probe) experiment, but the data was inconclusive. Thus, for more than a decade, the status of the Giant Void was the subject of bitter fights among scientists in the observational cosmology community.

In May 2009, while I was on sabbatical from UNC and living in England as a visiting professor at the University of Cambridge, I sat with a few dozen other scientists in a room at the Kavli Institute for Cosmology to watch the launch of the Planck satellite from the European Space Agency. The satellite's name was chosen to honor Max Planck, one of the founding fathers of quantum theory. Aboard the satellite was a powerful telescope designed to produce the most accurate measurements of the gentle glow of light left over from the fiery birth of our universe, the CMB. When the countdown began, the room fell eerily quiet. Liftoff was met with cheers and loud applause. Planck was on its way.

In March 2013, four years into its mission, the Planck satellite released the most finely detailed measurement of the CMB ever mapped. I was again in Cambridge, this time for a conference, listening to the press reports and preparing for bad news. Perhaps Planck's data and observations had ruled out all the anomalies in the sky that Rich, Tomo, and I had predicted in 2005 and 2006. Instead, just the opposite happened.

Planck's map contained a bombshell: The anomalies in our sky, including the Cold Spot, could not have been caused by anything in our own universe because they violated the uniform distribution of structure expected from cosmic inflation in a single universe. They had to have come from a different, noninflationary source outside its borders.

Later that day, I was scheduled to lecture at a conference. When the conference host introduced me, he remarked casually, "I believe Laura received some good news this morning."

It was good news indeed. As we had thought, the truth was sitting right there in our sky, waiting for us. It turned out that entanglement did actually leave imprints on our own sky, and these imprints are indeed strong enough to be detected by the technology we have today. Our universe, it seemed, was not alone. Nor was it such a fluke after all.

In addition to the Cold Spot, Rich, Tomo, and I predicted six other anomalies. Most of these anomalies, we calculated, would be strongest near the edge of our universe because that is where the net effect of entanglement remains the largest and purest. At smaller distances, we predicted, the violent, nonlinear processes of star and galaxy formation, such as swirling clouds of gas and the turbulence of collapsing matter, explosions, and ejections of material from exploding stars, were so powerful that they would wash out any signs of the weak signal from the entanglement contributions.

At the largest scales, we predicted another giant void, this time the size of one of our hemispheres as viewed from our location on Earth. Although it has the same origins as the first Giant Void, this one is larger in size but much weaker in strength. This second gargantuan void covers half the sky and creates small differences in the matter content between the two hemispheres. We anticipated it would show up on observed temperature maps of the sky as a slight difference, an asymmetry (or lopsidedness) between the average matter contents of the northern and southern hemi-spheres. And we predicted it could be found by studying the CMB.

CMB radiation waves are named after musical harmonics and are called multipoles. The largest CMB wave, the monopole, with a wavelength of about twice the size of the whole universe, would be equivalent to the fundamental or first harmonic; the CMB dipole would be equivalent to the second harmonic; the CMB quadrupole, roughly a fourth of the universe's size, would correspond to the fourth harmonic, and so on. A CMB wave with a wavelength about the size of the universe, the dipole distance, is what divides our sky into two hemispheres.

The effect of entanglement on our universe's gravitational potential is to (gently) deplete the long-wavelength CMB harmonics. This depletion manifests itself as a suppression of the CMB spectrum at the lowest harmonics, in addition to the slight hemispheric differences in the amount of matter content in the universe, and that is how we encountered the second giant, hemisphere-size void.

To understand why entanglement would deplete these CMB harmonics rather than amplify them, it is useful to go back to the analogy of Newton's apple. Like the moon, the planets, and all the stars in the universe pulling on the apple, the surviving infant universes on the quantum landscape multiverse pull on our universe. This pull, however weak, would still be significant and noticeable at the largest scales (the first few lowest harmonics), which is why the giant voids and other anomalies are found at those greater distances.

The hemispheric asymmetry in the CMB that we predicted was confirmed observationally by the Planck satellite in 2013 and 2015, two years after it found the first Cold Spot.*

---

* In 2018, I teamed up with a member of the Planck satellite experiment collaboration and a young, exceptional Italian astrophysicist, Eleonora di Valentino, to check the status of the predictions of my theory against the final release of the Planck data set. We found that the observations of all the anomalies agreed with the predictions.

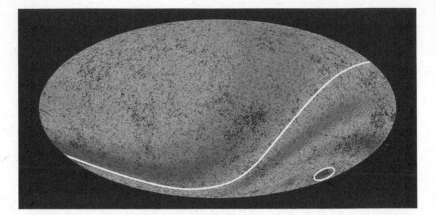

Figure 13. The CMB map observed by the European Space Agency Planck satellite indicating the Cold Spot (circled) and the asymmetry of matter between the two hemispheres (separated by the curving line). When the map was released in 2013, its surprising anomalies were in complete agreement with our predictions about the scars of the universe's birth that should be visible in our own sky.

Based on the fact that the remnants of entanglement are stronger and more easily observed at the largest distances, we made an additional prediction that the overall CMB temperature would be lower at those distances. Again, observational findings, including by the Planck satellite, show that the CMB temperature at the lowest harmonic scales (the size of the universe to its edges) is less than expected. And it's not just the temperature that is lower; the net effect of entanglement, at all the scales we estimated, is to decrease the overall strength (amplitude) of the inflationary CMB radiation (known as sigma 8) by around 20 percent. The combined observations of the CMB, including the Planck data, have also confirmed this prediction.

One of our other predicted signatures of the multiverse relates to the standard model of particles. The standard model of particles explains the origin of all known elementary particles in nature, from quarks to photons. Elementary particles are the offspring of a single fundamental particle—the Higgs particle, sometimes referred to as "the God particle." The Higgs particle was predicted in 1964 by Peter Higgs at UNC Chapel Hill, where he had been invited as a visiting scholar by Bryce DeWitt. The Higgs particle exists at energies of the order of a quarter of a trillion electron-volt, or tera-electron-volt (TeV). The particle was discovered in 2012 at the Large Hadron Collider (LHC) in Switzerland. The LHC is the world's largest machine; it is located in a circular underground tunnel with a circumference of twenty-seven kilometers. Beams of protons circle around the tunnel many times over before they collide, head-on at full speed, with each other. The energy of their collisions, which is a few TeV, is the highest energy we have produced on Earth.

Theoretically, the Higgs particle has a problem: its energy can grow in an uncontrolled manner and can destabilize the whole universe. In the 1970s, particle physicists postulated the existence of an underlying hidden symmetry of nature, known as super-symmetry, that would force the Higgs particle to remain stable. Once supersymmetry had done its job of protecting the Higgs particle at higher energies, it needed to be broken down at lower energies and ultimately disappear in order to allow the Higgs particle to decay into quarks and other particles. Protecting the Higgs particle with supersymmetry was an appealing theory, and the LHC was designed to test its existence.

In our quantum landscape multiverse theory, each wave-universe has its own individual, supersymmetry-breaking energy. This is because different wave-universes localize in different landscape energy sites and therefore start their Big Bang inflationary energies at different scales. But as we discovered, these wave-universes prefer to settle on high-energy sites of the landscape, which led us to conclude that if supersymmetry existed and its breaking was responsible for producing these high-energy landscape vacua, it would be bound to energies much higher than the anticipated Higgs energy. Therefore, we predicted that this supersymmetry breaking energy would not be discovered at the energy level of the Higgs particle when it was reproduced by the collider but rather at a billion or more times higher. And indeed, the Large Hadron Collider did not find supersymmetry at the Higgs energy level.

In less than a decade, six of our seven predictions had significant if not definitive proof to support them. Our seventh prediction, which involves the motion of galaxies relative to the expansion of the universe (known as "dark flow"), remains an open question. Observational results from two different teams were inconclusive, and a NASA-led team of astrophysicists that observed dark flow has not yet finished their project due to funding cuts. I hope they will be able to resume their work someday.

Far beyond what we hoped for when we started, our predictions of entanglement anomalies had been tested and observed. What's more, the Cold Spot observation was accurate at a sufficiently high confidence level to be considered a discovery. (In physics, the difference between an observational finding and a discovery is related to a statistical estimate of confidence level in the finding, called sigma. A sigma value higher than 4 or so indicates discovery because it means that the error in that particular observation is extremely low, and therefore the confidence level in the result is very high—what in daily parlance would be considered "beyond a shadow of a doubt." The Cold Spot observation had a sigma value of nearly 5.)

But does this mean we have complete proof for the origin of our universe from a quantum multiverse? No. Nature guards her secrets carefully!

One of the obstacles to moving from evidence to proof of the multiverse is not the squabbles over our models or the technology; it's a statistical problem known as cosmic variance.

Statistically, the more samples we have to measure, the more reliable our conclusions are. For example, if we measure a property of a galaxy—let's say its temperature—then the more (and similar) galaxies we can perform our measurement on, the more reliable our findings become. Suppose astrophysicists have a theory of galaxy formation that predicts their temperature should be about half a million degrees. If they have only one galaxy available to measure, and they discover that their prediction is correct and indeed the temperature is half a million degrees, then they are happy, but they are also aware that their finding could be a fluke. Statistically, if they have only one sample, their chance of being right is only about 50 percent. Who can say that if they had measured a second galaxy, they would find the same temperature? However, if they measure a trillion galaxies and find that most of their temperatures hover around half a million degrees, then the chances of error are pretty small—in fact, they are as low as one in a million. That error rate is so low that their finding can be considered conclusive; it becomes a discovery.

The trouble with cosmology is that, when it comes to searching for signatures that lie near the horizon of the universe, such as the lowest CMB harmonics, we have only one sample, one universe on which to make this measurement: ours. We cannot repeat our measurements in a trillion other universes. Despite sophisticated technology, we will always have a large statistical error rate at the largest distances because we can measure only one universe. This is cosmic variance. Therefore, although all the CMB experiments found these large-scale anomalies, this detection carries an inherently large statistical error. We cannot overcome the problem of cosmic variance by improving our technology; it is a statistical problem, one that will always be present because we have only one universe that we can measure.

However, the observational evidence in support of our six predictions establishes overwhelmingly that we are part of a multiverse. First, although it is possible to construct a model to explain *one* of these anomalies after it has been observed, constructing a model that not only explains but also predicts *all six* of them retroactively and simultaneously under one theory is nearly impossible. Even more important, making accurate predictions before observational results are known—rather than explaining the anomalies after they have been observed—should be both powerful and persuasive.

Second, recall that two overriding predictions of Big Bang inflation are uniformity and homogeneity. The significance of our observed anomalies is that they break this uniformity principle; they cannot be explained by a single, inflationary universe produced solely through cosmic inflation. They require a second source that additionally affected the formation of the CMB and all the structures in our universe—a presence that my collaborators and I argue, persuasively, I hope, is a quantum multiverse.

There are limits set by nature in our pursuit of proof and evidence for the multiverse. Owing to the speed-of-light limit, we cannot directly observe structures beyond the horizon of our universe, and we are also constrained by cosmic variance when measuring at distances near the horizon of the universe. Does this mean we should give up hope of ever deducing information about the multiverse? I don't think so. Consider this example: If you look at your arm, you can see neither the atoms in it nor the protons, neutrons, and electrons inside those atoms. If you are looking at yourself in the mirror, you can't see electrons flying around in your head as your neurons fire. However, your conscious thinking tells you not to doubt for a minute that you are made up of atoms. You don't question the truth of atoms in your body because you know the theory that has been tested in labs and in the sky: the standard model of cosmology in tandem with the standard model of particles. You know it provides a coherent answer for the whole chain of events that led to us. You wouldn't need to test the standard model of particles or cosmology on your body to know that it is correct.

Likewise, our theory of our universe being part of the quantum multiverse provides a consistent and coherent story of both our existence and what lies beyond, and it offers a series of predictions that are supported by all our observations. Our theory demonstrated that the answer to our origins can be derived, and using quantum entanglement, it proposed how to scientifically test the existence of the multiverse. And these reasons are sufficient to make me believe in the existence of a vaster, more complex, and more beautiful cosmos of which our universe is just a small part.

By demonstrating mathematically that the most likely way to start a universe is from high energies in a landscape of possibilities and by showing how to test its origins from a quantum multiverse, my work with Tomo and Rich stands in sharp contrast to previous estimates that gave our universe a nearly zero chance of existence. Instead, we could demonstrate that our universe is not at all special! I argued earlier in the book that cosmic inflation was an incomplete story of our universe because it could not explain its own origin. Our theory offers a completion of the standard model of cosmology by extending the cosmic story to the time before the Big Bang and to realms beyond our universe. It gives a coherent story that can be tracked step by step in the evolution of our universe, from its beginning as a quantum wave packet settled on some vacua on the landscape through its Big

Bang inflationary explosion and growth into a large classical universe bearing the scars of its origin from the quantum multiverse on its skies.

As the multiverse moved into the realm of scientific study, researchers became increasingly aware that a single-universe scenario was deeply problematic. Hints of the multiverse had been there all along, but they went unnoticed because of prejudice and focus on the theory of everything.

Today, this state of affairs seems to be changing. While I was writing this book, many scientists who once worked toward a theory of a single universe switched camps and are now working on models of the multiverse in an understanding of our origins as being simply a single chapter in a larger cosmic story. What was for years, indeed millennia, considered a radical idea is now mainstream. And despite the late Stephen Hawking's prediction that the theory of everything for a single universe would be discovered before the year 2000, he himself, like many of his peers, started working toward a multiverse theory in the twenty-first century.

Science is crossing the knowledge threshold to the moment of creation and the time before, and what we are finding is poised to upend centuries of cherished theories. We are at an unprecedented moment in scientific history because for the first time, the rules of nature and the origin of the cosmos are not simply a theoretical construct—they can be tested and proven. Indeed, this paradigm shift from a single universe to many moves science from the quest for a mega-theory to that of a multiverse; it extends the Copernican principle to the whole universe. If it passes all the tests, it will be one of the most important discoveries in the history of humanity.

11

Infinity and Eternity

IN THE YEARS following the debut of our multiverse theory, research on the multiverse has gone from the cosmological fringes to being a very active, popular field. My theory about the multiverse, once an outlier, is now far from the only one.

Interestingly, one of these alternative multiverse theories comes from Roger Penrose, the physicist who inspired my quest all those years ago and who we first met in this book as one of the strongest proponents of a singular universe explained by the theory of everything.

In 2007, three years after I proposed my theory of the multiverse, I invited Roger to UNC Chapel Hill. A giant void had just been observed, and I was excited by this dramatic development. I wanted to talk with Roger and get his reaction, especially because he has also been a world leader of the efforts toward a theory of everything. So, while I respected his opinion, I also expected resistance.

I collected Roger from the airport and drove him to his hotel on campus. He told me he would like to have dinner because he had been traveling all day. While I drove and listened to his travel stories, I was panicking inside, knowing full well that our chances of finding a late dinner in a small town like Chapel Hill were not high. The kitchen at his hotel was already closed, and the chef and staff had all gone home. Roger and I walked along the main road, Franklin Street, trying to find an open restaurant. After a mile or so, we spotted a Turkish restaurant called Talulla's. Its owner was just placing the Closed sign on the front door. I rushed over and pleaded with him to stay open, explaining that I had a very important guest and we had to find him something to eat. The restaurant owner's kindness and hospitality saved the day. Without hesitation, he welcomed us inside and apologized that he had only cold mezes; the stoves had been shut off. Roger and I took a table, and the owner brought over an array of dishes, then sat at the bar with a beer, patiently waiting for us to finish our dinner.

One of the most enjoyable aspects of discussing a hard problem with Roger, even when we disagree, is that his incredible passion for physics comes through; he is like a five-year-old in a toy store who is told he can have anything he wants. We discussed the recent landscape crisis, and I shared my opinion on the multiverse, then Roger excitedly told me about his own idea of a sequential universe.

In Roger's model, the beginning and the end of the universe are connected in a sequence. Roger envisions that in the distant future and thanks to dark energy, our universe will empty out completely; all observers will suffer the cosmic heat death, and our universe will keep a constant entropy forever. By the laws of physics, time will stop; we cannot build clocks in a universe where nothing changes. The universe is so unchanging—so big, uniform, smooth, and empty—that we can rescale its size from big to small without losing any information. (This rescaling of the size of the universe and everything in it is, in the language of math, "conformal transformation.") When we rescale, what we have in our hands is a small and smooth bit of space with lots of energy that we now know will again bang into inflated growth, go through the typical history of a universe, and then, in the far future, empty out, freeze, and start over.

In Roger's proposal, since the inflaton at the beginning and the dark energy at the end of the life of a universe are identified as one source, the universe keeps repeating these cycles, producing an infinite number of sequential universes or, as Roger calls them, aeons. Each aeon is a universe; therefore, a collection of universes spread out in time is also a multiverse. Interestingly, since clocks freeze at the end of each cycle, the second law of thermodynamics is not violated by Roger's theory because the arrow of time resets in each aeon. But it comes at the price of having a sequential multiverse, one where universes are produced one after another.

As we talked excitedly and wrote pictures and equations on paper napkins snatched from nearby empty tables, I kept an eye on the owner, who was still sitting at the bar, drinking his third beer, and talking on his phone in Turkish to his family while politely trying not to look at us as a signal to hurry up. But unhappily for him, we still had more to discuss.

Near the end of our dinner, I realized that, as much as Roger was motivated to find a unified theory of a single universe, his model inadvertently produced a multiverse; indeed, an infinite number of sequential universes. So it, too, produced a multiverse, but his collection of universes existed in time rather than space. I made this point and tried to convince Roger that all our attempts to solve the mystery of the origin of our universe would always end up with a multiverse—his model included. We debated this for hours and finally left the restaurant around one in the morning—but not before thanking the owner profusely.

My discussion with Roger that night at Talulla's remains one of the most intellectually memorable I have ever had. Roger and I have met and publicly debated the origin of the universe on numerous occasions since. In one of our more recent discussions, he announced that he had revised his "ridiculous number" concerning the likelihood of our universe coming into existence—he increased it from 10 to the 123rd power to 10 to the 124th!

Roger's multiverse evasion of the second law of thermodynamics is wholly original, but there are others that have rivaled it for sheer novelty. On a somewhat similar trajectory to Roger, Andrei Linde and Alan Guth, the founding fathers of cosmic inflation, have tried to solve the problem of our universe's origin from the day they realized their model of cosmic inflation produced the insurmountable problem of its origin, as Roger had demonstrated. In fact, Linde was the first one to propose a vision of an eternally reproducing universe, another type of sequential multiverse. Linde argued that if cosmic inflation can spontaneously happen once, then it may spontaneously do so over and over again. A single universe may multiply to produce new bubble universes that then branch out and similarly keep reproducing, creating yet more offspring. (Our four-dimensional universe is flat, as I've mentioned, but in its three spatial dimensions, it looks like a sphere. Thus the term *bubble universe,* which refers to our universe's three-dimensional shape rather than its four-dimensional flat geometry.) If we have eternity at our disposal, then we have all the time in the world to wait for more episodes of inflation to occur and produce new bubble universes. Each bubble universe bangs spontaneously from a patch of the previous one, which itself had inflated and branched out from a patch of its predecessor, and so on.

Linde's eternal inflation theory is an appealingly organic view of the multiverse. The closest comparison, indeed, is a natural one—just like a very old tree keeps growing new branches and leaves every year, this inflationary universe will keep endlessly reproducing new bubble universes.

But Linde and Guth's eternal-inflation theory suffers from a familiar problem: it closes off the origins of the universe from examination, just as Penrose and Hawking's singularity theorem did when it posited that nothing, absolutely nothing, existed before creation. Under eternal-inflation theory, reconstructing our universe's origin from its parent universes all the way down their genealogical tree is like trying to trace the origin of a single leaf all the way to the first shoot from the seed of an infinitely old and tall tree on which the leaf appeared. An eternally reproducing universe would take us from bubble universe to bubble universe, all branching out from each other, endlessly back into the past. If eternal inflation is correct, then our origin is pushed all the way back to the infinite past, hidden under an infinite number of origins of all the other bubble universes that came before it. In short, our beginnings become untraceable because the first moment of creation is hidden in the eternal past.

The inflationary-universe community has recently looked for ways to test eternal inflation. They do so by resorting to something even spookier than the quantum entanglement that I used as a testing mechanism in my theory. In eternal inflation, bubble universes, being continually produced, end up colliding with each other. Thus, tests of eternal inflation rely on the assumption that collisions between our universe and other bubble universes would create "fractures"—another sort of imprint on our sky. Of course, in the majority of these cases, collisions of two bubble universes would be catastrophic, and we wouldn't be here to talk about them. But according to eternal-inflation colleagues, it was conceivable to anthropically "arrange" for the collisions of our universe with other bubbles to be so soft that they didn't destroy our universe.

I'm not sold on those ideas. Observationally, if two spheres or two bubble universes collide softly with each other, based on their symmetry, we would expect to see the ripples they produce on their respective surfaces; they would look like concentric circles spreading out from the point of impact. These types of ripples have not yet been observed in our sky.

Still, my curiosity got the best of me. In 2013, together with Malcolm Perry, a colleague of mine at the Department of Applied Mathematics and Theoretical Physics at the University of Cambridge, I decided to undertake my own effort to understand eternal inflation. Specifically, we wanted to investigate whether eternal inflation was indeed eternal and if it could be unified with the theory that Tomo, Rich, and I had presented about our universe originating from the quantum landscape multiverse. If so, then we could arrive at a unique image, a unified theory of the multiverse.

Eternal-inflation theory relies on the random quantum fluctuations of our old friend the inflaton particle. Sometimes these fluctuations can cause the inflaton to spontaneously jump in energy, creating energies high enough to trigger a local Big Bang and therefore produce a bubble universe. But having an inflaton jump to high energies is not sufficient to produce a bubble universe. For an inflaton to produce a universe, the high-energy fluctuations also need to find an incredibly smooth, tiny region of space, a piece of prime real estate on which to start this bubble universe.

In turn, the production of multiple bubble universes makes the background space coarser and coarser, so finding smooth prime real estate becomes harder, and it becomes more and more difficult, if not impossible, to produce additional bubble universes. A good analogy is an ice-skater trying to skate on ice that is becoming covered with dirt and particles; when the ice is no longer smooth, the skater cannot continue fluidly skating. Our skater will be forced to come to an abrupt stop due to the friction produced by the contact of the skate's sharp blade with the gritty buildup on the ice's surface. Similarly, the universe's reproduction becomes harder to continue if the space-time is increasingly "choppy." Malcolm and I found that in this scenario, eternal inflation eventually ceases. Building a multiverse, apparently, is harder than it seems.

Through frequent discussions with my late friend and colleague Stephen Hawking, I could observe how fast the scientific thinking about the multiverse was changing. Hawking, who spent most of his life working toward a theory of everything and was one of its iconic leaders, initially embraced the view of an anthropic selection of our universe from the string-theory landscape. Gradually he shifted away from the theory of everything and became open to the possibility of a multiverse by investigating eternal inflation. Yet once he had convinced himself that an eternal-inflation model was also problematic (which he did, independently of Malcolm and me and for different reasons), he started exploring a new model. In the last few years of his life, Hawking was actively working on the physics of the multiverse.

Hawking's conversion on multiverse theory was indicative of the about-face that the field of physics was undergoing as the new millennium progressed. What had once been a fringe idea was now firmly in the mainstream.

To be fair, in the early years of working on the multiverse, I was not a complete anomaly. Another independent advocate for the multiverse paradigm was MIT's distinguished theoretical physicist Max Tegmark; he proposed that many of the deep mysteries in physics, from dark energy to the existence of life, could be better explained if our universe was not alone. Tegmark advocated for a mathematical multiverse—a multiverse where all the possible mathematical objects, seen and unseen, from, say, a doughnut-shaped universe to a flying spaghetti monster, acquired a physical existence. It is a fascinating view based on entropy arguments with far-reaching philosophical implications, but the mathematical-multiverse theory is harder to test observationally than our quantum landscape multiverse theory.

Tegmark's theory and those of Hawking, Guth and Linde, and Penrose are only the most popular models of the multiverse; today, there exist many others. But what all these theories have in common is that they posit something that sixty years ago was considered unthinkable at best and heresy at worst—the idea that we do not need a theory of everything for our singular universe; instead, we exist in a multiverse.

A few years ago, I was debating this shift at the annual HowTheLightGetsIn festival at Hay-on-Wye in Great Britain with a conservative opponent of the multiverse. Exasperated by the arguments in favor of the multiverse, he declared to me and the audience: "But half of the physics community does not believe in a multiverse!" He paused briefly in order to let his grave statement settle in, and to lighten the mood, I replied jokingly: "So you are agreeing that the other half of the community *does* believe in the existence of the multiverse?" Everyone laughed. And in my heart, I knew that we were both right.

The German philosopher Arthur Schopenhauer put it best when he said, "Truth passes through three stages. First, it is ridiculed. Second, it is violently opposed. Third, it is accepted as being self-evident." Today, many scientists view as self-evident the possibility of our cosmos being vaster than a single universe.

For the first time, we have what we need to look up in the sky from our little planet and see and test the far reaches of cosmic theory, beyond the horizon of our one universe. Now, from inside our finite and ephemeral universe, we can finally reach for infinity and eternity.

Epilogue: A Place to Dream

In scientific history, including that of the twenty-first century, discoveries are often made by parting company with established theories—despite a scientist's best intentions not to. After all, breaking from the mainstream isn't easy; established theories are what a scientist grows up with, is inspired by, and cannot do without. They are often a scientist's lifelong partner and best friend. At the same time, discoveries are a scientist's own brainchild; they are the enticement that attracted her to science in the first place and the legacy that survives her. The lucky ones have one or two breakthrough ideas during their lifetimes.

Often, though, the exaltation of a breakthrough idea is tested against the desire to maintain loyalty to established theories. This test happened to the founders of quantum theory, Planck, Einstein, Bohr, Heisenberg, and Schrödinger, and other great scientists of their generation. Confronted with this particular test, at least, many of these great minds failed—at first.

In such situations, scientists face a dilemma: they have to choose between their own theories and the theories that were the foundations on which their work was built. It is an impossible choice.

Difficult though that choice may be, however, any scientists worth their salt will choose the same way: they will select the path of investigation, of knowledge, of testability, no matter how steep the climb may be. It was no different for the founding fathers of quantum theory. At first, they resisted breaking away from the determinism of classical physics. But once convinced that they had the correct answer, they did the unthinkable. Scientific integrity overruled rigid austerity. They gave in and courageously stood tall against the might of classical physics. They bravely crossed over to the quantum realm and forever changed the way humans think about our world.

Perhaps the theory of the multiverse is poised to cause another, similar paradigm shift—forever altering how we conceive of our world and our place in it. We know that our universe is not eternal, and it is not infinitely large. It had a beginning 13.8 billion years ago, and it has grown to about 10^{27} centimeters—its present size. These are big numbers but certainly not inconceivable for minds like ours. We have every right to wonder what existed in the cosmos fifteen billion years ago and what the cosmos looks like at 10^{32} centimeters, beyond the horizon of our universe. We have every right to wonder—and investigate.

Cosmology is not a new field. In fact, it is among humanity's oldest intellectual endeavors. All traditions of ancient mythology contain tales of the origin of the universe, sometimes based on gods and supernatural forces, other times on observations of the night sky combined with critical thinking. Human beings were asking probing questions about the cosmos long before telescopes, computers, Einstein's equations, or quantum theory existed.

We cannot plot our future without a map of our past, and our past has some very surprising lessons for our present. Many of the models and theories of the universe in modern times can trace their roots to the wisdom and ideas of the ancients. And many of today's battles over whether our universe is the center of the cosmos or just one in a vastness of universes have been fought before.

My introduction to the evolution of scientific thought came from my father. After reading Russian translations of English books, my father translated them and read them back to me in Albanian. Every Sunday that he was not in exile, my father shared his knowledge. My more practical mother sent her two intellectual dreamers to the National Library, which had a café and cake shop on the terrace and three floors of archives below.

To this day, I can still smell the green vinyl covering of the floors, see the spiral staircase rail I used to slide down, and hear my dad delivering seminars on the history of scientific thought to his devoted and attentive audience of one. I loved the stories he told about philosophers' lives and work, great thinkers who debated and fought, sometimes even coming to blows, over cosmic issues. He was the person who introduced me to all the ways to see the proverbial forest despite the thousands of individual trees. My first exposure to prohibited Western literature and the evolution of ideas was through his translations.

The roots of Western cosmological thought can be traced all the way back to ancient Greece. No matter how far science believes it has advanced, the leading schools of Greek philosophy still echo in our present-day perceptions of the organization of our universe and what lies beyond. In many ways, our most modern ideas about the universe began there. Around 400 BCE, the philosopher Democritus, who was born into a wealthy and powerful family, used his wealth to travel to India, Egypt, and throughout the Mediterranean to absorb knowledge from other cultures and scholars. From his mentor Leucippus, he adopted the idea that the world was made up of indivisible clumps of matter (atoms) and empty space (voids) through which those atoms moved. Democritus also believed in a deterministic universe in which all events could be estimated and anticipated with 100 percent certainty. In Democritus's world, the motion of atoms was mechanical. It followed a set of rules and was completely predictable.

The Democritean model of a universe's creation starts with a collection of atoms moving around the voids, then clumping together to form larger objects such as stars, planets, and the whole universe. Because there are an infinite number of atoms and voids, this process can be continuously repeated to form many universes, each of which meets its end when it collides with another and then is atomized back into individual particles. Thus, Democritus was the first recorded Western philosopher to imply that our world might be part of a multiverse.

Plato, a student of Socrates and perhaps the most influential thinker in Western science and philosophy, was a contemporary of Democritus, and it is not an exaggeration to say that Plato hated his fellow philosopher. It is alleged that Plato so loathed Democritean ideas that he wanted to burn all Democritus's writings. I was astonished to read that Isaac Newton, two thousand years later, did the same with all the writings and portraits of fellow scientist and predecessor at the Royal Society Robert Hooke in an attempt to wipe out every trace of him. This is a recurring theme; many scientific conflicts have been repeated through three thousand years of history with the same viciousness, dedication, and passion as in our own modern times. (Though perhaps with a little less venom in the pre-internet era.)

While the Democritean model describes the material world in terms of atoms and voids, Plato took another position. He posited two levels of existence: the physical world that we see around us and an abstract, higher plane of existence, the form or demiurge, which fashions and maintains the physical universe.

Plato's student Aristotle, the forefather of the natural sciences, argued for a different view; he believed that the cause of an event could be found in the physical world, that there was no need for a higher power or demiurge. In order to explain the motion of planets and stars without a demiurge, he embraced the view that the universe was a collection of etheric rotating celestial spheres to which planets and stars were "glued," with the Earth fixed at the center of the spheres. Aristotle's ether was an invisible, crystal-like material that filled space. His universe had no beginning and no end; it was eternal but limited (spatially) to about twenty thousand times the size of the Earth.

His model was supported by the geocentric Ptolemaic system developed by the Alexandrian astronomer Claudius Ptolemy in the second century CE. In the geocentric system, Earth sits at the center of the universe. This Ptolemaic view of planetary motion occurring through rotating celestial spheres dominated astronomy for almost two millennia, until the time of Galileo.

But Aristotle, Plato, and Democritus were not the only ones to have sweeping ideas about our universe. I think it is stunning that in the third century BCE, the Greek atomist philosopher Epicurus reasoned that there was a universal "uncertainty principle"—more than two thousand years before it was postulated by Heisenberg! Epicurus wanted to allow for human free will. He argued that if the course of atoms was predetermined and predictable, then people's lifetime experiences had to be predetermined, since people were also made of atoms. But, he went on, if that were true, then people simply become bystanders to their own existence rather than participants in it. Which could not be.

Consequently, rather than a Democritean, predictable, deterministic world, Epicurus advocated for an unpredictable—indeterministic—universe. He postulated that nature allowed for random small deviations or the swerving of atoms in the voids, which moved them away from their predetermined course. Collectively, these deviations in groups of atoms, whether in humans or the whole universe, made their course undetermined, so they could be spoken of only in terms of chances. Two incredible facts about Epicurus's life stand out. First, his reasoning in favor of an uncertainty principle and an indeterministic universe was reached through arguments for the existence of free will within human morality, yet it ultimately led him to establish the quantum uncertainty principle. Second, many of his original writings were preserved by a catastrophic accident—a library of papyri of his work was discovered inside the residence of a wealthy Roman senator during excavations of the town of Herculaneum in the eighteenth century. The papyri had been carbonized and preserved by volcanic ash from Mount Vesuvius's eruption in 79 CE.

The importance of Epicurus's work in physics and ethics cannot be overstressed. Thomas Jefferson considered himself an Epicurean, writing, "If I had time, I would add to my little book the Greek, Latin and French texts, in columns side by side, and I wish I could subjoin a translation of Gassendi's *Syntagma* of the doctrines of Epicurus, which, notwithstanding the calumnies of the Stoics and caricatures of Cicero, is the most rational system remaining of the philosophy of the ancients, as frugal of vicious indulgence, and fruitful of virtue as the hyperbolical extravagances of his rival sects."

The philosophy of Epicurus was reintroduced to the Roman world and Western philosophy by the poet Lucretius in his epic poem *De Rerum Natura* ("On the Nature of Things"). Unlike Plato, Epicurus rejected the need for divine intervention in the formation and regulation of the universe. Like Democritus, he thought of the world as an infinite space and time containing many universes—a multiverse. But unlike Democritus, Epicurus believed the world had no certainty.

The fundamental argument between a Democritean deterministic versus an Epicurean indeterministic world is much the same one that continues to this day. The two sides have remained in a tug-of-war, with both camps for decades divided between an Einstein-inspired classical universe, where every event can be estimated and determined, and a competing quantum multiverse theory of an indeterministic cosmos that allows for many universes.

Before they reached our own era, these ancient Greek theories traveled across the intervening centuries. In the thirteenth century, the Aristotelean view of the cosmos fascinated the Western theologian and philosopher Thomas Aquinas. Aquinas tried to reconcile the Aristotelean paradigm with Christianity. While still respecting Aristotle's view, Aquinas posited a universe that had a beginning and therefore required divine intervention to come into existence.

In early-sixteenth-century Poland, the astronomer Nicolaus Copernicus, in a paper published just before his death in 1543, formulated the heliocentric system, in which the Earth was one of a collection of planets orbiting the sun, even though for much of his lifetime, Copernicus was reluctant to break with Aristotelean authority and continued to subscribe to Aristotle's celestial spheres.

A few decades later, in the 1570s and 1580s, the Danish astronomer Tycho Brahe found that the planets, including the Earth, were indeed orbiting the sun. Using detailed measurements, he showed that the paths of comets passed around the sun and therefore through the celestial spheres. For the first time, there was observational evidence to challenge Aristotle's shell model of the universe. German mathematician and astronomer Johannes Kepler (1571–1630) drew upon Brahe's calculations to produce the laws of planetary motion, confirming the sun's central position in our solar system.

Final proof came from one man: Galileo Galilei, born in Pisa in 1564. Galileo learned of the newly invented telescopes produced by spectacle-makers in Holland. In 1608 and 1609, he built his own telescope to conduct astronomical observations. Applying his telescope and his genius, Galileo formally confirmed the existence of the heliocentric system, ensuring that the Earth would never again be viewed as the center of the universe. (Along with this new proof of how the universe was organized came efforts to calculate the age of the universe. In 1650, James Ussher, an erudite and influential Irish Anglican bishop, declared that the universe had started at 6:00 p.m. on October 22, 4004 BCE. This declaration may seem strange today, but his ideas were taken seriously until subsequent geological research in the nineteenth and twentieth centuries found that fossils of "terrible lizards," or dinosaurs, that once walked the Earth were millions of years older.)

Aristotle's views of the universe were finally laid to rest by Isaac Newton, an incontestable giant of physics and, as the astronomer royal Lord Martin Rees said, "the best student Cambridge University ever had." In his book *Philosophiae Naturalis Principia Mathematica* ("Mathematical Principals of Natural Philosophy"), first published in 1687, Newton gave the world a theory of gravity and motion that showed that the heliocentric system was held together by gravity. To this day, Newton's theory continues to provide explanations for the motion of most objects in the universe and on Earth.

But Newton firmly and, as Einstein was to demonstrate later, incorrectly, believed that space and time were the absolute building blocks of the fabric of the universe—that is, they were always there, like a permanent container, a sort of bucket inside which everything else moved and existed. Newton's theory of gravity dominated physics until the late nineteenth century, when further discoveries in mathematics and physics led scientists to challenge that model.

Newton was convinced space and time were a structure that was always there, supporting the universe and the motion of celestial bodies within it. Newtonian space and time would appear like a rigid board, one that could not change its shape no matter how much weight was put on it. Einstein saw things differently—and it profoundly changed our understanding of the nature of our universe.

Einstein's theory of relativity transcended Newton's theory (just as Newton's theory had transcended Aristotle's—such is the history of science). It gives a universe where nothing exists before the singularity; that is, a universe without an absolute clockmaker. A universe where every event except creation can be calculated and predicted with certainty.

Up to this point, geometry had been based on the work of the Greek mathematician Euclid, who published his *Elements* around 300 BCE. Euclid had deduced the principles of his geometry from a small set of axioms, starting with "The shortest distance between two points is a straight line," an axiom we can easily grasp. Indeed, we know that if we were to walk from, say, the Art Institute to the Wrigley Building in Chicago, the shortest path would not be a zigzag or a circle but a straight line along Michigan Avenue.

The problem is that Euclidean geometry breaks down when applied in a curved space-time, such as the space-time of our universe. To see why, imagine for a moment that the universe looks like the surface of a sphere, roughly like planet Earth. On this sphere universe, we want to travel from Chicago to Tokyo. We can see on the globe that the shortest path joining Chicago and Tokyo is part of a circle that goes over the North Pole. It is definitely not a straight line. The reason that the shortest path is an arc is due to the curvature of the sphere. If we had applied Euclid's straight-line geometry to the curved space-time of our example, we would have been misled.

In the late nineteenth century, mathematical breakthroughs, notably with the work of Riemann, Lobachevsky, and Minkowski in non-Euclidean geometry, began to transform physics and established the foundation on which a new theory of the universe could be built. The beginning of the twentieth century witnessed the greatest revolution thus far in the history of physics. Not unlike Plato's ancient view of two levels of existence, twentieth-century physics also found two levels of existence: the macroscopic, visible world, a single, deterministic universe governed by Einstein's theory of relativity; and the microscopic, unseen world, inhabited by atoms, electrons, particles, and waves, whose workings were captured by quantum mechanics. But the most disruptive element of a quantum universe is not its small size; it is its indeterminism, which means that every event in it, including its own creation, is uncertain and based on probabilities.

And that is where we began our story in this book. We have seen how a quantum universe based on probabilities allows for the existence of many worlds—in other words, the multiverse.

The journey from antiquity to the third millennium may seem long. But put in perspective, it took about 3.8 billion years for life to emerge on Earth. In another four billion years, our

Milky Way galaxy is set to collide with Andromeda, a nearby galaxy, an event that will likely wipe out our planet. Even before that event, our planet will be too hot from the increase in the sun's luminosity, and life as we know it will be extinct in about one billion years.

Viewed at this scale, five thousand years are a mere blink of an eye. Yet in that blink of an eye, humanity, through imagination, observation, and courage, has journeyed to the very edge of the universe and the first millisecond of its conception 13.8 billion years ago and to a theory of the creation of our universe—an achievement of astounding proportions. Today, through the power of physics, observation, supposition, and mathematical proof, we can now reach back to the moment before our universe was conceived.

Truly, today we have gained the ability to travel beyond the confines of our own universe, if only in our minds. But perhaps in the process, we have discovered something more important: that our universe and our very existence arose from a bizarre quantum-probability game and that our universe is but a humble member in an intricate, vast, and breathtakingly beautiful cosmic family.

When I was growing up, for two weeks every summer, my parents rented a holiday apartment by the beach in their hometown of Vlora, an ancient coastal city along the Adriatic Sea. Known as Aulona in Greek and Roman times, it remained a special place to visit even during 1980s Communist Albania. Aulona's spirit, imprinted on the traditions, superstitions, and landscape of the place, floats outside of time. The town is guarded by a rugged terrain of high mountains, turquoise waters, and black rocks that blend into silence at sunset. It is a place to dream.

My favorite evening activity during these family vacations was to sit on the sand alone. I would watch the waves linger at the soundless horizon and then break rhythmically onto the shore. As night fell, I waited until the line dividing sky and sea blurred away and all boundaries vanished. Of course, everybody knew that the world beyond the horizon was strictly forbidden to those of us behind the Iron Curtain. But sitting in the dark, I was free to imagine. Were the kids who lived on the other side of the Adriatic, in Italy, equally enchanted by the edge of the sky and sea we shared?

Eventually my dad would come over and, without reprimand, sit on the sand next to me. Then it was the two of us in a hushed conversation with the sky. Before long, he would tell me it was time to leave, and the gentle spell of the sea and the sky would break.

In 2013, twenty years after our family's last trip to Vlora and the year the Planck satellite was launched, I returned to my favorite spot in Vlora with my three-year-old daughter. She was excited to be there, completely carefree, happy to splash sand and water in every direction. How stunning, I thought—how far we'd come in just a generation, politically and scientifically. My daughter belongs to a generation and a country that does not have to accept limits on imagination and discovery. She can live by the motto that my father taught me: "Without knowledge, existence is in vain."

And how much further we have come in the brief time since this trip. Consider for a moment the possibility of a far more complex and richer cosmos made up of many universes—a multiverse in which our universe is but a single, humble member in a far-flung corner of this vastness. That possibility allows for the mathematical calculation of a range of values for habitation; it allows us to objectively compare chances of existence of different universes, rather than the logically flawed comparison of a single universe to itself, and it allows us to derive and explain our origins, not postulate them on anthropic grounds. It offers a glimpse of the cosmos beyond our horizon and before the Big Bang. And rather than shutting the door on scientific inquiry, it pushes us to think more broadly and boldly.

The scientific knowledge that the human race has accumulated so far is a glorious chapter in the book of our species' endeavors. But the next frontier in the relationship of humans with nature—laws above and the multiverse below—is waiting to be written. When we take up the pen, we will be bound by nothing— nothing, that is, save for the limits of our own imagination.

Acknowledgments

This is it. I have crossed the finish line, written the book, and I haven't done it alone. I would like to thank the many friends, colleagues, and family members who contributed to the completion of this project.

First, thank you to my scientific collaborators, some of whom I have already mentioned, for sharing the ups and downs of this journey of discovery. I am also grateful to Professor Christian Iliadis, my colleague and friend at UNC, for his enthusiastic support.

A special acknowledgment goes to Peter and Amy Bernstein, my dear friends and agents at the Bernstein Literary Agency. Peter and Amy, thank you for your guidance, encouragement, and direct help with editing the whole manuscript multiple times, something very few people would do. And thank you for your continuous trust in me throughout the years.

I am grateful to Alexander Littlefield, executive editor of Mariner Books at HarperCollins, who mentored me through the writing process and managed to make me do something I haven't done before—talk about myself! His high professional standards are demanding, but at the same time the incredible care, scrutiny, and thoroughness he invested in every line of the book have been very helpful and motivational during the crucial final editing stages. I would also like to express my gratitude to Stuart Williams, executive editor, and his deputy Will Hammond at Bodley Head, Vintage Penguin, who worked closely with Alex to provide invaluable editorial support.

Many thanks go to Lyric Winik, who patiently helped me edit many versions of the book as it evolved into its final form.

I would like to acknowledge the help of many friends who read earlier drafts and provided feedback and support: Phil Doran and Nick Ward in Cambridge, England; Rob Westermann and David Ballinger in the United States; and Dhurata Sinani in Canada.

The support of my family has been incredible. I appreciate the help of my husband and companion on my journey, Jeff Houghton, who was the first person to read, candidly criticize when called for, and edit every line I wrote while providing backup on all the family fronts when I was busy writing. Special thanks also to my brother, Aurel Mersini; his son (my nephew), Dominic; and my mother, Stela Mersini, for their unwavering and uplifting support and their honest comments.

Last but not least, I would like to thank the two most influential people in my life, the ones to whom this book is dedicated and who, through their unconditional love and support, do not ever allow me to grow old:

My exceptional daughter, Grace Houghton, fills every day of my life with pride and happiness. Although she is still a child, her infectious optimism and maturity beyond her years, her love, and her encouragement are my inspiration. Grace, as you used to tell me when you were younger, I love you from here to infinity, with a love bigger than the multiverse!

Anyone reading my book will not be surprised to hear that the biggest influence on my earlier years was my wonderful father, best friend, and confidant, Nexhat Mersini, who died in 2011. Dad, I miss your quiet strength, your wisdom, kindness, integrity, friendship. I most of all miss our long conversations over triple espressos about any interesting topic, be it in science, math, arts, poetry, philosophy, evolution of ideas, or music. I hope I have done justice to your memory through our stories. Thanks, Dad.

Index